PRAISE FOR TOXIC POLITICS

"Dr. Huang has animated a dry subject and brought it to life. He elegantly combines environmental science, public health, political science, organizational behavior, and his own family and hometown experiences. The writing style is approachable and yet also very careful and precise. The book is a tremendous read. It is a worthy successor or update of Liz Economy's *The River Runs Black* of 2004. Moreover, this is much more than a scholarly work. The subject and analysis elucidate the conditions that at least 1.4 billion Chinese citizens face every day. The spillover effects of the problem-set that he so eloquently unpacks here will have obvious global consequence. I only hope that the book is translated into many languages. But, most importantly, I hope that it is translated into Chinese and read widely by members of the Chinese public as well as the political establishment in China."

– Craig Allen, President, US Ambassador (ret),
US–China Business Council

"This book addresses a key question as to how a major environmental health crisis may influence China's rise. The book systematically reviews the fundamental economic, sociopolitical, foreign policy and health implications of the crisis and its policy response. The book also offers an intriguing assessment of the evolvement and implementation of environmental health policies in China. It is a must-read for those who are interested in global environmental health and policy issues."

– Xi Chen, Associate Professor at Yale University; President,
China Health Policy and Management Society

"Bringing well-researched data to life through poignant personal narratives, Huang documents devastating air, water, and soil pollution – and the Party's conflicted approach to acknowledging and addressing its toll on citizens' health. To understand some of

China's most pressing challenges now and in the future, read this book!"

– Dr. Andrew S. Erickson, Professor, Naval War College; Visiting Scholar, Harvard Fairbank Center

"*Toxic Politics* does not hedge or equivocate. China must first heal its body politic before it will effectively address its environmental slide and heal its people. Willing participation by citizens in environmental and health action, as well as in overall governance, is necessary for a healthy environment and a sustainable great power."

– David M. Lampton, Oksenberg-Rohlen Fellow at the Asia-Pacific Research Center, Stanford University, and former director of China Studies, Johns Hopkins –SAIS

"Yanzhong Huang has accomplished a remarkable achievement: he offers us a comprehensive, truly interdisciplinary study of interrelated Chinese political, economic, and societal dynamics through a deep dive into the relationship between two vitally important policy areas, environmental degradation and its effects on citizens' health."

– Andrew Mertha, George and Sadie Hyman Professor and Director of China Studies, Johns Hopkins University

TOXIC POLITICS

Environmental degradation in China has not only brought a wider range of diseases and other health consequences than previously understood, it has also taken a heavy toll on Chinese society, the economy, and the legitimacy of the Party-state. In *Toxic Politics*, Yanzhong Huang presents new evidence of China's deepening health crisis and challenges the widespread view that China is winning the war on pollution. Although government leaders are learning, stricter and more centralized policy enforcement measures have not been able to substantially reduce pollution or improve public health. Huang connects this failure to pathologies inherent in the institutional structure of the Chinese Party-state, which embeds conflicting incentives for officials and limits the capacity of the state to deliver public goods. *Toxic Politics* reveals a political system that is remarkably resilient but fundamentally flawed. Huang examines the implications for China's future, both domestically and internationally.

Yanzhong Huang is a senior fellow for global health at the Council on Foreign Relations. He is also a professor at the School of Diplomacy and International Relations at Seton Hall University, where he developed the first academic concentration among US schools of international affairs to explicitly address the security and foreign policy aspects of global health issues. His writing has appeared in outlets including *Foreign Affairs, Foreign Policy*, the *New York Times*, and the *Washington Post*.

A COUNCIL ON FOREIGN RELATIONS BOOK

The Council on Foreign Relations (CFR) is an independent, nonpartisan membership organization, think tank, and publisher dedicated to being a resource for its members, government officials, business executives, journalists, educators and students, civic and religious leaders, and other interested citizens in order to help them better understand the world and the foreign policy choices facing the United States and other countries. Founded in 1921, CFR carries out its mission by maintaining a diverse membership, including special programs to promote interest and develop expertise in the next generation of foreign policy leaders; convening meetings at its headquarters in New York and in Washington, DC, and other cities where senior government officials, members of Congress, global leaders, and prominent thinkers come together with CFR members to discuss and debate major international issues; supporting the Studies Program that fosters independent research, enabling CFR scholars to produce articles, reports, and books and hold roundtables that analyze foreign policy issues and make concrete policy recommendations; publishing *Foreign Affairs*, the preeminent journal of international affairs and US foreign policy; sponsoring Independent Task Forces that produce reports with both findings and policy prescriptions on the most important foreign policy topics; and providing up-to-date information and analysis about world events and American foreign policy on its website, www.cfr.org.

The CFR takes no institutional positions on policy issues and has no affiliation with the US government. All views expressed in its publications and on its website are the sole responsibility of the author or authors.

TOXIC POLITICS

China's Environmental Health Crisis and Its Challenge to the Chinese State

A Council on Foreign Relations Book

Yanzhong Huang

Council on Foreign Relations and Seton Hall University

CAMBRIDGE
UNIVERSITY PRESS

University Printing House, Cambridge CB2 8BS, United Kingdom

One Liberty Plaza, 20th Floor, New York, NY 10006, USA

477 Williamstown Road, Port Melbourne, VIC 3207, Australia

314–321, 3rd Floor, Plot 3, Splendor Forum, Jasola District Centre, New Delhi – 110025, India

79 Anson Road, #06–04/06, Singapore 079906

Cambridge University Press is part of the University of Cambridge.

It furthers the University's mission by disseminating knowledge in the pursuit of education, learning, and research at the highest international levels of excellence.

www.cambridge.org
Information on this title: www.cambridge.org/9781108841917
DOI: 10.1017/9781108895132

First published 2020

A catalogue record for this publication is available from the British Library.

ISBN 978-1-108-84191-7 Hardback
ISBN 978-1-108-81528-4 Paperback

To my father, Huang Yujin,
and to the memory of my mother,
Yan Weiying (1939–2016)

Contents

Figures

Tables

Acknowledgments

When I started preparing this book five years ago, I set the goal of completing the project in two years, without knowing that eventually it would be the events and developments in the environmental and political worlds that drove the book-writing process. Venturing into uncharted territory, I have been privileged to receive encouragement and support from a lot of people. They include my colleagues at Seton Hall University's School of Diplomacy and International Relations: Courtney Smith, Andrea Bartoli, Benjamin Goldfrank, and Martin Edwards, who gave me considerable latitude and support when I was working on the book.

I have benefited immensely from the help and support of many of my current and former colleagues at the Council on Foreign Relations: Alyssa Ayres, Tom Bollyky, Liz Economy, Laurie Garrett, Josh Kurlantzick, Shannon O'Neil, Adam Segal, Varun Sivaram, Sheila Smith, and Benn Steil. As a leading voice on China and director of CFR's Asia Studies Program, Liz has offered invaluable advice and support from the very beginning of the book project. CFR Director of Studies James Lindsay also deserves a special mention because he patiently read all versions of the project proposal and manuscript and offered extensive advice and suggestions, from which I benefited enormously. CFR President Richard Haass took the time to read the manuscript carefully, and his feedback pushed me to make the arguments more focused and coherent. I would also like to extend my profound gratitude to Patricia Dorff, who shepherded me through the book publication process. Amy Baker, Dominic Bocci, Jean-Michel Oriol, and Shira Schwartz also provided important support for the project.

I am grateful to the outside reviewers for their comments and suggestions on the book manuscript and Sara Doskow for her flexibility and guidance in publishing the book. Jonathan Landreth and Tom Redburn read through the entire manuscript in different stages of the project and provided excellent help in making this book accessible to a broader readership.

Throughout the process of research and writing, I was fortunate to receive research and editorial assistance from many graduate students at Seton Hall and young professionals at CFR: Erik Crouch, Lauren Greenwood, Kirk Lancaster, Gabriella Meltzer, Alex Miller, Ariella Rotenberg, Ben Silliman, Aaron Steinberg, Meagan Torello, and Dylan Yalbir. Gabriela Hasaj assembled the various components of the book, and Ebelle Ong Wei Shan provided critical support in data analysis.

I would also like to thank the Smith Richardson Foundation for the generous financial support and Allan Song for guiding me through the grant-seeking process. The funding enabled me to conduct fieldwork research and organize a series of meetings on China's environmental health. Through a CFR study group chaired by Jeffrey Sachs, I was able to collect constructive feedback on earlier versions of the book's chapters from a number of scholars and experts, including Frances Beinecke, Dan Fox, Mark Frazier, Joan Kaufman, Steve Morse, Steve Orlins, Deliang Tang, Yang Gonghuan, and Yu Zhou.

I also owe a great debt to the following friends, scholars, officials, and activists who facilitated the book's preparation and writing: Jan Berris, Chen Ajiang, Dingding Chen, Xi Chen, Deng Guosheng, Dong Dong, Feng Yongfeng, Jennifer Holdaway, Hu Tao, Mike Lampton, Ken Lieberthal, Liguo Lin, Lican Liu, Ma Daoming, Miao Lu, Minxin Pei, Pu Xiaoyu, Shang Qi, Judith Shapiro, Susan Shirk, Henry Wang, Jock Whittlesey, Xue Lan, Dali Yang, and Jack Zhang. Numerous other friends, scholars and officials in China helped arrange my research and fieldwork in Beijing, Jiangsu, Shanghai, Shanxi, and Guangdong during various stages. Out of respect for their anonymity, I am unable to acknowledge their help individually.

Born into a family in rural China, I dared not dream of going to college, let alone the opportunity to travel across the ocean and find a home in a beautiful and bountiful land. I am deeply indebted to my parents, who not only taught me honesty and integrity in life but also showed their understanding and support of my pursuit of an academic career. My mother and my first teacher, Yan Weiying, passed away while I was working on the book. The experience of my father, Huang Yujin, who has worked hard all his life but still regrets not being able to provide a decent life for his children, reminds me not to lose touch with the less well-off in my research. I dedicate this book to my father and the memory of my mother.

Abbreviations

$\mu g/m^3$	micrograms per cubic meter
ADB	Asian Development Bank
AHC	adjusted human capital
API	Air Pollution Index
AQSIQ	General Administration of Quality Supervision, Inspection, and Quarantine
AQI	Air Quality Index (instead of API)
ARGs	antibiotic-resistance genes
BLL	blood lead levels
BTH	Beijing–Tianjin–Hebei
CBCGDF	China Biodiversity Conservation and Green Development Foundation
CCG	Center for China and Globalization
CCP	Chinese Communist Party
CCTV	China Central Television
CFR	Council on Foreign Relations
CEPI	central environmental protection inspection
China CDC	China Center for Disease Control and Prevention
CPPCC	Chinese People's Political Consultative Conference
CNPC	China National Petroleum Corp
CO_2	carbon dioxide
COP21	2015 United Nations Climate Change Conference
COPD	chronic obstructive pulmonary disease
CVDs	cardiovascular diseases
DALYs	disability-adjusted life years
EIA	environmental impact assessment

EPA	US Environmental Protection Agency
EPBs	Environmental Protection Bureaus
EPI	Environmental Performance Index
EPIL	environmental public interest litigation
EST	Environmental Supervision Talks
FGD	flue gas desulfurization
GBD	Global Burden of Disease
GW	gigawatts
HDI	Human Development Index
HEI	Health Effects Institute
IHME	Institute for Health Metrics and Evaluation
IOC	International Olympic Committee
IPE	Institute of Public and Environmental Affairs
LNG	liquefied natural gas
MEE	Ministry of Ecology and Environment (2018 successor of the MEP)
MEP	Ministry of Environmental Protection (the successor of SEPA)
MFA	Ministry of Foreign Affairs
MIIT	Ministry of Industry and Information Technology
MLR	Ministry of Land and Resources
MOF	Ministry of Finance
MOA	Ministry of Agriculture
MOFCOM	Ministry of Commerce
MOH	Ministry of Health
MOHURD	Ministry of Housing and Urban-Rural Development
MOST	Ministry of Science and Technology
MOT	Ministry of Transport
MSW	municipal solid waste
MWR	Ministry of Water Resources
NCDs	noncommunicable diseases
NDRC	National Development and Reform Commission
NEA	National Energy Administration
NEMS	National Environment Monitoring Station
NGO	nongovernmental organization
NHFPC	National Health and Family Planning Commission

Nimby	"not in my backyard"
NO_2	nitrogen dioxide
NO_x	nitrogen oxide
NPC	National People's Congress
NSFC	National Natural Science Foundation of China
O_3	ozone
PDP	pollutant discharge permit
PM	particulate matter
PPP	public–private partnerships
PRT	pollution rights trading
PX	paraxylene
REE	rare-earth elements
RSCs	regional supervision centers
SARS	severe acute respiratory syndrome
SASAC	State-owned Assets Supervision and Administration Commission
SEPA	State Environmental Protection Administration (successor of State Environmental Protection Agency)
SEPA	State Environmental Protection Agency
SFDA	State Food and Drug Administration
Sinopec	China Petroleum & Chemical Corporation
SO_2	sulfur dioxide
SOA	State Oceanic Administration
TEC	total emission control
TVEs	township and village enterprises
UNCHE	UN Conference on the Human Environment
UNEP	United Nations Environment Program
VOCs	volatile organic compounds
VSL	value of a statistical life
WHO	World Health Organization
WTE	waste-to-energy

Introduction

AS THE NEARLY 3,000 DELEGATES TO THE NATIONAL PEOPLE'S Congress (NPC), China's legislature, arrived in Beijing in March 2013, they found the annual political fanfare overshadowed by appalling air pollution in the country's capital. Only days before the meeting, a measure of air quality that tracks health-threatening fine particles had reached a hazardous level more than ten times the level ever recorded in Los Angeles, long considered one of the most smog-prone cities in the United States.

Dubbed the "airpocalypse," the dark cloud over China's legislature was not a momentary lapse or a problem confined to Beijing. In December 2014, when the high-speed train I took barreled into the city of Baoding in Hebei Province, I felt as though I had arrived in a ghost town, as everything outside the window was shrouded by an oppressive and toxic smog. The haze did not fade even after the sun came out. Forty-five minutes later, the train pulled into Shijiazhuang, the provincial capital – another ghost town.

Baoding and Shijiazhuang are by no means outliers when it comes to air pollution in China. Beginning in mid-December 2016 and through the New Year holiday, a wave of dangerous and dirty smog extended from northeast China to the south. The toxic smog blanketed a third of China's cities and in some places was so bad it went off the standard scale. The choking smog surrounding the city of Kaifeng, in Henan province, was so heavy that there were reports of birds fainting from hunger after low visibility made it impossible for them to catch their prey.[1] The smog forced Beijing and many more provinces and cities to issue their first red alert of the year, the highest alert in a four-tiered pollution warning system. Schools were shut down, flights and factories were suspended, and cars were barred from the roads. Sensing huge demand for fresh air, a Canadian company did a brisk business selling air bottled in the Rocky Mountains town of Banff for $14 to $20 per canister.[2]

Polluted skies are not the only environmental catastrophe in China. Water and soil pollution are now part of the grim new normal in China as well. A bulletin issued by the Ministry of Environmental Protection (MEP) in 2015 suggested that more than one-third of China's surface water and two-thirds of its underground water supplies were polluted.[3] About one-fifth of Chinese people do not have access to safe drinking water.[4] While the issue of water pollution has been rife for many years,[5] soil contamination became a major concern only recently. The public got a glimpse of the scale of the problem when a 2014 government report admitted that as much as one-fifth of China's farmland soil is severely polluted with toxic chemicals.

Such environmental problems pose a serious and sustained threat to the health and well-being of the Chinese people. Here environmental health problems are defined by Phil Brown as "health effects caused by toxic substances in people's immediate or proximate surroundings."[6] This definition focuses on the direct pathological effects of air pollution, water pollution, and soil contamination. It avoids addressing in detail hygiene-related or vector-borne disease threats (e.g., chronic diarrhea, schistosomiasis), social determinants of health (e.g., housing, urban development, and land use), or environmental hazards that do not have a direct impact on health, such as desertification, deforestation, and acid rain.

China is certainly not the only country that is adversely affected by environmental health challenges. The health effects of environmental degradation in advanced industrialized countries, including Japan, the United States, and Great Britain, are well documented.[7] But the scope of China's environmental health crisis has reached another dimension, raising critical questions about the ability of the Chinese state to cope with its complex domestic and international challenges.

Not only do air, water, and soil pollution plague China with a wider range of diseases and public health problems than previously thought, but they are also taking a heavy toll on China's society, economy, and polity. The health and non-health consequences of pollution, as well as the complexities involved in addressing the problem, have generated an environmental crisis that will test the Chinese state in unprecedented ways. China's evolving response to environmental health challenges suggests that government leaders are learning and responding; indeed they have demonstrated the ability to introduce new policy tools and transmit pressures down to the lowest level of the hierarchy to get things done. Nevertheless, the state commitment to pollution control is undercut by the gaps and deficiencies in policymaking as well as the perverse bureaucratic incentives created in the implementation stage.

An analysis of both qualitative and quantitative data challenges the widely held belief that China is winning the environmental health battle, showing that the stricter and more centralized policy enforcement measures introduced by Xi Jinping, China's maximum leader since 2012, have not produced truly significant and substantial gains against pollution. As a result, the regime confronts a profound quandary between the need to sustain rapid economic growth and the costs of environmental cleanup that will challenge its ability to maintain political legitimacy. An examination of China's environmental health governance highlights an authoritarian system that is remarkably resilient but fundamentally flawed, which has profound implications for the viability of the so-called China model.

Furthermore, the analysis highlights the linkage between China's environmental health problem and its international ascendance. In September 2002, in recognition of the lack of understanding in Washington of China's internal challenges, a think tank hosted a conference entitled "China in Transition: A Look Behind the Scenes." Bates Gill, a leading China expert, set the stage: "The China we will face in five to ten years' time will be fundamentally shaped by the outcomes of the dramatic political, economic, and social transformations unfolding in that country today. The better we understand these transitions today, the better off we will be in engaging China tomorrow."[8]

Since then, a voluminous literature has addressed China's internal challenges – economic, political, social, and environmental.[9] Scholars and pundits alike agree that such domestic problems may constrain

China's international rise, but few studies thus far have sought to provide a cogent and coherent analysis of the linkage between the two. In fact, most studies on China's rising power tend to highlight China's economic prowess and military buildup, as well as its ability to project its international influence, but fail to pay adequate attention to another important dimension of state power: the state capacity to extract and mobilize resources, to provide public goods and services, to have its claim to rule willingly accepted, and to enforce rules and regulations across its entire territory. How a major internal challenge is translating into stumbling blocks in China's rise can be shown by examining the foreign policy implications of China's environmental health crisis and the state's capacity to effectively respond to the crisis.

OBSTACLE TO GLOBAL LEADERSHIP

Many of China's environmental health problems today harken back to the pre-reform era, when Mao Zedong's development strategy had tragic consequences for human beings and the environment.[10] The current crisis, however, is unfolding in a new socioeconomic and political context. The leaders of post-Mao China have been remarkably effective in laying the foundation for stunning gains in the economic well-being of hundreds of millions of people. But the conflict between economic development and environmental protection is exacerbated because the performance-based political legitimacy on which the Chinese Communist Party (CCP) relies has led to a single-minded pursuit of economic growth to the detriment of the environment and the health of the people.[11] Moreover, after four decades of rapid industrialization and modernization, the sheer size of China's economy and population coupled with its growing integration into the world economy multiplies the challenges to the country's health, environment, and society.[12]

The changing nature and magnitude of the crisis has tremendous economic, sociopolitical, and foreign policy implications for China's reemergence on the world stage. For most of the last two millennia, China was one of the dominant cultural, economic, and political actors in the world.[13] It came to be seen as a feeble power only after its humiliating defeat in a succession of wars with foreign powers in the nineteenth century. Emboldened by four decades of robust economic growth, Chinese leaders have been promoting the "China Dream," which is about realizing a prosperous and strong nation, rejuvenating its greatness in the world. As President Donald Trump has abdicated the United States' vital leadership role and decided to retreat into what he calls an "America First" policy,[14] China has moved to fill the void left in global governance, defined as the rules, structures, and processes that guide and regulate the international system and life among people of different countries. China's emerging global leadership has been welcomed by a number of world leaders. UN Secretary-General António Guterres told Xi that it was "very reassuring to see China assuming such a clear leadership in multilateralism in today's world."[15] "The world needs a leading force like China," International Monetary Fund Managing Director Christine Lagarde said in April 2018.[16]

Yet China's international ascendance is neither linear nor inexorable. It faces several "stressor variables," such as entrenched corruption, a widening wealth gap, and a deteriorating environment. Pollution alone may take a heavy toll on China's economy, society, and even national security. For example, Wu Shuangzhan, former commander of the Chinese People's Armed Police, said that years of pollution in a rural town of a southern Chinese province were so severe that none of the individuals who sought to enlist in the military over the past nineteen years met the basic physical requirements.[17] Wu's concern about pollution's effect on national security recalled Murray Feshbach and Alfred Friendly Jr.'s documentation of "ecocide" and environmental degradation in the former Soviet Union.[18] It is hard to imagine that China can regain its greatness in the world if the Chinese people do not have clean air to breathe, safe water to drink, or uncontaminated soil on which to live and farm.

Environmental health problems involve a deeply moral question in terms of the relationship between pollution and affected populations. Here the focus is not as much about environmental soundness as it is about the health consequences of pollution, which carry with them the prospects of deep grief among individuals and communities and convey a sense of gravity against a looming threat. As the health and social costs of environmental degradation go up, public discontent over environmental health issues are also on the rise, putting the country's sociopolitical stability at stake. Some leading Chinese experts have noted that if the government fails to address the crisis effectively, the expanding gap between economic and social development – combined with growing social frustration over a worsening environment – could devolve into a much bigger crisis of legitimacy.[19] Their forecasts echo what Jared Diamond described in *Collapse: How Societies Choose to Fall or Succeed*, in which environmental damage is found to be a major factor contributing to the collapse of societies, from Easter Island to the Maya of Central America.[20]

Environmental health challenges also add directly to the legitimacy problems for China's global leadership. Under the Belt and Road Initiative, a grand vision to deepen physical infrastructure, financial, political, and security connections in Eurasia, China is lending billions of dollars to participating countries for coal-fired power-generation projects while giving short shrift to renewable and green energy. Many see this move as an attempt to shift China's pollution problem to its neighbors.[21]

ENVIRONMENTAL HEALTH CHALLENGES

Societies suffering from environmental damage are not doomed to failure. Very often, a crisis response leads government leaders to undertake extraordinary steps to push for the introduction of a new and broad policy agenda, to develop new institutions for structuring the often clashing interests and interactions between the multitudes of policy actors involved, and to make sure society and the bureaucracy are mobilized effectively and efficiently. For China, the environmental crisis is likely to require profound political changes beyond routine policymaking and implementation. As Elizabeth Economy has noted: "Turning the environmental situation in China around will require something far more difficult than setting targets and spending money; it will require revolutionary bottom-up political and economic reforms."[22]

Compared to other crises, environmental health problems have some unique features that warrant special attention. First, the environment–health nexus can be notoriously complex. Given the wide range of hazards individuals may be exposed to, establishing a cause–effect relationship between specific environmental risks (e.g., polluted water from a factory) and health problems (e.g., cancer deaths in a nearby village) is extremely difficult.[23] This complex causality provokes tensions and conflicts over responsibility, not just between victims and polluters, but also between jurisdictions and regions, calling for a multifaceted government response.[24] Second, unlike many traditional sociopolitical problems, pollution puts everyone's health at risk. Even though the effects are not felt equally in different communities, or even by each person within a community, they affect rich and poor, privileged and destitute, powerful and powerless alike. The indiscriminate nature of the health concerns can be used to facilitate consensus-building and broaden the constituency calling for change.[25] The scale and seriousness of the challenges, though, can narrow the state's options in containing social protests, especially when the protests involve whole communities with a common grievance.

Third, pollution's health effects differ, depending on what pollutants are involved and where and when they emerge as threats to health. "Risks to health from food vary widely from product to product and present very different kinds and levels of threat to human health from sources including heavy metal contamination, pesticides, and veterinary drugs," Jennifer Holdaway and Wang Wuyi have noted. "The same is true of different types of air and water pollution, which have different composition, sources, and impacts, from short term acute exposure to long term cumulative effects through climate change and the degradation of ecosystems."[26] Because of regional diversity and uneven economic development, different localities are under the sway of different mixes of environmental health problems. For example, coal burning is a more serious threat to health in Beijing than in Shanghai, where there is no wintertime coal-fired central heating.[27] In contrast to Beijing, haze is not a major concern in Guangzhou, where emissions from the petrochemical industry is the primary environmental health problem.[28] Smog composition varies from city to city even in the same province.[29] All this requires the state to adopt targeted but comprehensive policy interventions.

Finally, environmental health issues do not fall under the institutional responsibilities of a particular government agency. Instead, addressing the challenges requires the involvement of a whole range of government agencies: environment, health, land use, urban planning, and so on. With multiple government agencies competing for jurisdiction, tensions are bred among these agencies due to conflicts of interest and contestations over responsibilities.[30] This is further complicated by the challenge of coordination among subnational governments. Pollution as a negative externality is often regionally produced; regional transport of pollutants, for example, contributes up to 70 percent of the smog in Beijing.[31] Regions impacted by smog do not match the boundaries of administrative jurisdictions, making it important that provincial governments cooperate in order to address pollution effectively. Here provincial government leaders face a typical "collective action" problem in smog control: they would be better off collaborating, but none of them wants to be the single one taking actions (and thus bear all the costs of smog control) with no guarantee of cleaner air.

RESPONSE FROM SOCIETY

If environmental health problems have been exacerbated by the post-Mao reform, the social response to the crisis is embedded in a political milieu dramatically different from the Mao era. Modern reforms resulted in the "individualization" of Chinese society, as characterized by the untying of the farmers from the collective, of the economy from central planning, and of the individual from her indigenous community.[32] This process, coupled with fiscal and bureaucratic decentralization, led to the emergence of a more complex plurality of interests that span across rural, urban, and virtual spheres. In the field of environmental health, not only are there different types of hazards (air, soil, and water pollution) that occupy different temporalities (anticipated pollution, "outbreak" events, and "slow disasters"), but the types of social action also vary from isolated and ad hoc to more organized and confrontational collective action.[33]

This has transpired at a time when post-Mao reform has ushered in new opportunities for society to protect itself against the intrusive reach of the state. Widespread Internet use, for example, has enabled Chinese citizens to engage in practices ranging from diffuse dissent to radical protests.[34] Post-Mao state-rebuilding has also promised to increase the space where non-governmental and civil society organizations operate.[35] Their participation in policymaking can potentially be facilitated by various state-led institutional arrangements that serve to channel social demands into the state policymaking regime.[36] For example, the Environmental Impact Assessment Law, unveiled in 2002, requires public hearings for major development projects. While the law is rarely enforced, people now have numerous methods through which to address issues that affect their health and livelihoods, including petitioning the local environmental protection bureaus, alerting provincial media about the infractions of local factories, and piggybacking on other grievances such as land rights abuse.[37] As a result, environmental health concerns have become one of the primary reasons behind social complaints and confrontational collective actions. By 2016, according to a Pew report, 70 percent of the public viewed air pollution as a "very big" or "moderately big problem," up 7 percentage points from 2008.[38]

The waxing power of society, however, does not necessarily mean that the power of the state is waning in the policy process. Compared to liberal democracies, an authoritarian state boasts more policy arenas that are not subject to institutionalized negotiations between state and society; this autonomy bestows the state the advantage in policymaking and enforcement. Such "despotic power"[39] can be magnified when state elites are managing perceived crises.[40]A large literature suggests that central and local state actors in China developed new sources of power vis-à-vis society by reconstituting the levers of central control[41] and, for local state actors, by actively promoting industries in their localities.[42] Research by Lora-Wainwright also shows that despite the increasing resources and capabilities at their disposal, Chinese citizens' strategies and actions, as well as the very way they conceptualize environmental health risks, are shaped by their fear of state repression, state-defined opportunity structures, and dependence on polluting industries.[43] Indeed, while farmers worry about the health risks from polluting local factories, many industrial workers in the same community downplay the risks because of their financial dependence on the same factories.[44] Similarly, in the absence of trust in the government's ability to ensure food safety, people may reduce exposure to toxic elements in food by simply shopping more carefully rather than publicly demanding the state tighten food safety regulation.

Since the economically disadvantaged's choices over what they can purchase and consume are limited, economic disparities have led to uneven access to safe food.[45] In this sense, the impact of pollution on health may illuminate people's "deep ambivalence about development and modernization and some of the new fault lines of inequality and social conflict that they generate."[46] Furthermore, such individualization of risk reduces the likelihood of collective action, fragmenting popular protests against pollution in China into more sporadic and localized actions.

THE STATE RESPONSE

Until the late 1990s, the government pushed aside environmental considerations in pursuit of short-term breakneck growth, leading to catastrophic environmental damage.[47] In the beginning of this century, the environment–health nexus began to draw attention from top officials with the publication of a policy report by two Chinese think tanks.[48] Since then, the issue has taken on increasing urgency, highlighted by a series of reports from the government, international organizations, and think tanks.[49] Still, until fairly recently, there were almost no systematic data on the health effects of environmental degradation. It was not until 2010, for example, that the now widely used measures of PM2.5 – fine particles in the air that are smaller than 2.5 micrometers in diameter and capable of causing serious heart and lung problems at high concentrations – became part of the official lexicon. In the absence of reliable and systematic indicators, shocking and sudden "focusing events" became the only catalyst to have problems recognized, redefined, and addressed formally. Yet until 2012, events focusing attention on environmental health were rare, not only because there was a lack of awareness, but also because environmental health crises are essentially "slow disasters" that do not generate as immediate or dramatic impact as "outbreak events" such as severe acute respiratory syndrome (SARS).

With the opening of China to the global society, international actors have played a significant role in changing the priorities of policymakers, moving latent public health issues on to the government agenda and affecting the timing of government action and content of policy design.[50]

International agencies and foreign governments have been instrumental in engaging, socializing, incentivizing, and constraining domestic policy makers through the various means and resources at their disposal: reputation and pressures, information and ideas, and expertise and experience. In the run up to the 2008 Beijing Olympics, which served as a prime "coming out" party for Beijing on the world stage, Chinese leaders ratcheted up their rhetoric regarding the need to protect the country's environment. Meanwhile, the event prompted the US mission to monitor China's air quality level, especially the PM2.5 concentration, sending credible signals about the seriousness of the problem. With the help of social media, the extremely high PM2.5 levels in Beijing became common knowledge. The increased awareness of the environmental health problems triggered widespread public outcry that the government should treat pollution as a policy priority and issue action plans to address air, water, and soil pollution.

Of course, it is one thing to put a policy blueprint by central political leaders in place; it is another to have the local bureaucracy carry out the policy faithfully. A major problem Beijing has to grapple with is that the party-state does not have sufficient power to elicit compliance from lower levels.[51] As in electoral democracies, authoritarian leaders in China ("the principal") face three major challenges in implementing policy when dealing with local bureaucratic officials ("the agents"). The first is information disadvantage: principals in Beijing have limited capacity to observe the agents' actions and characteristics. The second is about leadership cohesiveness: the leaders themselves may disagree with each other in the policymaking process. The third is the limited range and credibility of the tools at their disposal for rewarding and punishing the agents' behavior.[52]

This principal–agent problem generated difficulties in policy implementation even in the Mao era.[53] Nevertheless, the concern was mitigated by "bandwagon politics," that is, because the distribution of political power was sharply skewed toward Mao, policy actors had strong incentives to jump onto Mao's policy bandwagon in order to be rewarded as early and enthusiastic supporters.[54] As a result, stalemate or foot-dragging was relatively rare. The institution of collective leadership at the top level in the post-Mao era made a single or unified principal less likely and the vast bureaucratic expansion created exceedingly complex and cumbersome lines of crisscrossing authority that makes direct communication and cooperation between functional units under different ministries a daunting task. Policy implementation also was confounded by the presence of a territorially defined chain-of-command. Fiscal and administrative reform decentralized control, giving rise to a "de facto federalism," or "a system of multiple centers of powers in which the central and local governments have the broad authority to enact policies of their own choice."[55] The redistribution of political and administrative power, in conjunction with the alternative sources of power and authority produced by the market-oriented economic reforms, gave rise to a much more diffuse power structure than in the Mao era.

The power shift was accompanied by other changes. The relaxed social and political control and the reduced use of purges and labeling against bureaucratic officials reduced the fear factor and limited the range and credibility of leaders' rewards and punishments. Rather than acting upon whatever the political leaders said, subordinates had strong incentives to exercise strategic disobedience in the policy process. As a result, "*every* central initiative [was] distorted in favor of the organization or locality responsible for implementation."[56] Moreover, local administrators, subject to myriad, often conflicting, demands from a number of superiors, tended to emphasize only a few objectives and ignore the rest. To make matters even worse, the post-Mao era saw the rise of a "buck-passing polity" under which local officials shirked their responsibilities in policy enforcement.[57]

To overcome the agency problems, Chinese leaders historically have relied on intense politically charged campaigns. By setting a clear goal with a measurable outcome and by mobilizing the society and bureaucracy to fulfill the policy objectives, the campaigns offered an ad hoc, stopgap solution in order to align bureaucratic behavior patterns with leaders' policy preferences. But such campaigns rarely led to the employment of the best set of technologies or incentives for behavioral change.[58] And once they achieve the stated target of completion, the leadership's attention dissipates, and society is back to business as usual.

Mass campaigns faded in the post-Mao era when central leaders moved some contentious policy issues from the political arena to the administrative "neutral" zone while seeking to use market forces as an alternative policy instrument. Instead, a new approach called cadre evaluation (*ganbu pingjia*) was introduced. This approach evaluates local party and government officials based on performance targets (*kaohe zhibiao*) designed by their immediate superiors, and the evaluation results are used to determine local officials' career path. Performance targets are ranked in terms of their importance. In contrast to "soft or ordinary targets" like healthcare, "hard targets" like GDP growth and "priority targets with veto power" such as population control were what could really make or break local cadres' careers.[59]

The cadre evaluation system provided strong incentives for local implementers to mobilize all needed resources to meet priority targets identified by their bosses. Yet, because local officials care only about how they are scored, the system also created moral hazards by encouraging local cadres to focus on issues that are quantifiable and/or identified as a priority by their superiors, such as local economic growth.[60] Moreover, it encouraged local officials to fabricate data in order to make themselves look good, which renders it hard for policy makers to receive effective feedback for possible policy adjustment. In recognition of the weaknesses of the system, China has further classified all indicators into two types: expected targets and binding targets. Expected goals, such as GDP growth, are desirable but failure to fulfill them will not necessarily invite punishment. Binding targets such as emission control level, on the other hand, must be fulfilled as prescribed by the upper levels of government. The central government aims to use the former to encourage the local government to sustain natural economic development, and the latter to modify local governments' incentives so that they respond to central requirements in areas such as providing public services or improving environmental protection.[61]

But ever since 2012, the dynamics at the top have changed again, as Xi Jinping, turning away from the collective leadership under Jiang Zemin and Hu Jintao, has rapidly centralized power and sought to recreate the Maoist bandwagon. Xi now seems to call the shots. The unprecedented anti-corruption campaign Xi launched in 2012 and the consequent purge of even top officials makes everybody in the power hierarchy feel insecure. This "Xi in Command" structure has generated strong incentives for government officials to compete to jump onto Xi's policy bandwagon and actively seek out his cues in the policy arena. With Xi's political and policy vision now permeating the power hierarchy, willful and wanton disobedience in the policy process becomes less likely.

But even as Xi has made environmental protection one of his highest priorities, China faces serious obstacles to actually implementing environmental health-related policies. While their policy rhetoric might suggest otherwise, Chinese leaders continue to be torn by the conflict between economic development and environmental protection. In order to ensure employment and avoid the so-called middle-income trap (getting stuck at a level of development that falls short of that of more advanced economies) the state must push for a relatively high level of economic growth. But under the existing energy and industry structure, that is prone to causing yet more pollution. And efforts to combat pollution, at least in the short run, undercut efforts to sustain GDP growth. The inability to clearly prioritize environmental protection diminishes policy cohesiveness which, in turn, creates room for policy manipulation by local cadres to the detriment of environmental health. Moreover, the reliance on the highest authority as the measure of all things (*dingyu yizun*) under Xi stifles creative thinking among bureaucrats in problem-solving and discourages them from taking bold policy initiatives. Bureaucratic inertia can be also triggered by the terror-inducing anti-corruption campaign, which reportedly has contributed to a "wait and see" attitude among bureaucrats tasked with its implementation.[62]

To be sure, the government has put in place a new set of policy tools for enforcing pollution controls. These policy measures carry the hope for bold changes, but their effectiveness also risks being compromised by the decentralized and crisscrossing policy structure, an underdeveloped market society, a weak environmental regulation regime, and increasingly tightened political and social control. The mixed results of existing policy measures are corroborated by an analysis of quantitative and qualitative data on pollution control. Official reports suggest that China has made tremendous progress in tackling the environmental health crisis. But a more in-depth analysis makes clear that the policy effectiveness continues to be procedurally incoherent, regionally uneven and temporally unsteady, not to mention the numerous unintended repercussions created in the policy process. In this sense, China now appears to be "lurching between accelerating environmental damage and accelerating environmental protection."[63] Policy effectiveness will remain limited unless the government commits to upgrading its toolbox, which would require profound changes in state–market relations, bureaucratic power structure, and state–society relations.

AUTHORITARIAN RESILIENCE

Given the sociopolitical implications of China's environmental health crisis and the governance requirements for crisis management, discussing China's environmental health governance fits squarely within the intellectual and policy debate over the Chinese state's capacity to revamp itself and the prospect for China's political development and global leadership. The debate originally unfolded against the backdrop of a global surge in democracy that began in the late 1970s and continued through the early 1990s.[64] The democracy movement gained additional momentum in the early 2000s, with the spread of "color revolutions" in former Soviet republics and the Balkans, and later, the uprisings against authoritarian rule in the Middle East and North Africa.

But today, three decades after the fall of Soviet communism, the euphoria has given way to pessimism. Not only has the number of representative governments fallen, but the quality of democracy has deteriorated in many countries as well.[65] With growing populism and increasing signs of polarization and intolerance, now even long-established democracies in the United States and Europe are experiencing the decay of democratic values, a process Larry Diamond called "deconsolidation."[66] In October 2017, twenty of America's top political scientists gathered at Yale to discuss American democracy, and nearly everyone agreed that the nation is eroding socially, culturally, and economically: "If current trends continue for another 20 or 30 years, democracy will be toast," one of the scholars concluded.[67]

As liberal democracies retreat, authoritarianism is rising in established autocracies (e.g., Russia) and nominally democratic countries (e.g., Turkey). One of the leading contributors to this worldwide setback for democracy is China.[68] The authoritarian regime not only squelched the 1989 democracy movement but also maintained astounding economic growth in the decades since. Impressed by the endurance of the communist regime and the adaptability with which the Chinese leadership handled internal and external challenges, many China watchers began to characterize the authoritarian political system as "resilient" and "strong."[69] In a special issue published by the *Journal of Democracy* in January 2003, Andrew Nathan outlined four important institutional changes in the Chinese political system that serve to consolidate authoritarianism: (1) the increasingly norm-bound nature of its succession politics; (2) the growing meritocratic considerations in the promotion of political elites; (3) the differentiation and functional specialization of institutions within the regime; and (4) the increasingly institutionalized political participation that strengthens CCP's legitimacy base.[70] Disguised in terms such as the "Beijing Consensus" and "China model," the authoritarian resilience theory soon became the mainstream explanation of China's economic growth and political sustainability.[71]

To be sure, the authoritarian resilience theory had its own critique from the very beginning. In the same special issue of the *Journal of Democracy*, Bruce Gilley cast doubt on Nathan's argument by pointing to the historical limits of institutionalizing the processes of elite promotion, the elite functional responsibility, and popular participation.[72] Interestingly, over time, some of the leading proponents of authoritarian resilience have flipped their positions. Nathan apparently changed his mind in 2009 when he wrote in an article entitled "Authoritarian Impermanence" that "[t]he most likely form of transition for China remains the model of Tiananmen."[73]

The leadership transition that began in 2012 rekindled a vigorous discussion in the China-watching community: does Xi Jinping's rise augur the rejuvenation of the one-party state or the beginning of its end?[74] The dramatic downfall of some of China's most powerful political actors highlighted major flaws in China's political system, including rampant corruption, sustained factionalism, nepotism and patronage in the selection and promotion process, elite contempt for the law, and the cutthroat nature of succession struggles.[75] At odds with his earlier idea of a resilient authoritarianism with capacity for learning and adaptation,

David Shambaugh, in a well-known *Wall Street Journal* op-ed article in 2015, expressed his pessimism about China's future: "[Xi Jinping's] despotism is severely stressing China's system and society – and bringing it closer to a breaking point . . . Its demise is likely to be protracted, messy and violent."[76]

The larger question at the heart of the debate is whether the so-called China model, which features single-party rule and state capitalism, will emerge as a successful rival to liberal democracy. Unlike other authoritarian states (e.g., Putin's Russia and Venezuela in the Chavez and post-Chavez era) China has been able to sustain relatively strong economic growth even though it is becoming politically more oppressive than in the early 1980s, when the reform was kicked off. Both Minxin Pei's analysis of "developmental autocracy" and Yasheng Huang's study on "capitalism with Chinese characteristics" highlight the party's insistence on dominating society and the economy as the biggest obstacle to sustainable economic growth and stability.[77] Pei further argues that beneath China's façade of ever-expanding prosperity and power is a Leninist state that is in its advanced stage of decay.[78] Joshua Kurlantzick, on the other hand, disagrees. His research on state capitalism points to the emergence of a powerful new rival model, even though he is concerned about its long-term effectiveness.[79] His view is echoed by Mark Thompson, who suggests that authoritarianism and economic growth might be reinforcing.[80] Convinced that China's model presents a viable, sustainable alternative to liberal democracy, some China-based intellectuals went a step further. In a 2013 *Foreign Affairs* article, Eric Li, a Shanghai-based venture capitalist, declared that CCP will not just stay in power; its success in the coming years will "consolidate the one-party model and, in the process, challenge the West's conventional wisdom about political development."[81] Daniel Bell, a professor at Tsinghua University, contends that China has developed a model of "democratic meritocracy" that is morally desirable and politically stable and can help remedy the key flaws of Western democracy.[82]

The mixed performance record under President Xi Jinping has not resolved the debate; if anything it has reinforced the beliefs on both sides. Economic and political reforms have clearly suffered serious setbacks in China. Since coming to power in late 2012, Xi has re-centralized political power, reversed the market-oriented economic liberalization, regained the party-state control in Chinese social-political life, and rejected Western concepts of "constitutional democracy," "civil society," and "universal values."[83] At the same time, China under Xi has sustained economic growth, stressed the importance of rule of law, solidified a campaign against corruption in China's officialdom, strived for pro-environmental and pro-health policies, and strengthened China's influence on the global stage.[84]

At a conference in New York, Yu Keping, a leading Chinese intellectual most well-known for his book *Democracy is a Good Thing*, argued that even though the formal political institutions have remained unchanged in China, there were clearly signs of "modernization of state governance."[85] While efforts to strengthen ideological control might reflect the leadership's fear that the regime is under serious threat, they may also suggest renewed confidence in the current system. Rising economic and military power and the growing problems faced by Western democracies (e.g., partisanship and political gridlock) have emboldened Chinese leaders and intellectuals to proclaim "firm confidence" in the China model. Taking advantage of the erosion of confidence in US democracy, the CCP, through its official news agency, published a commentary on the eve of the 19th Party Congress in October 2017 presenting the Chinese political system as a more credible alternative to crisis-ridden Western democracies.[86] More recently, Xi Jinping described China's authoritarian one-party system as "a new option for other countries and nations who want to speed up their development."[87] As if to show China is serious about exporting its governance model, the CCP held a World Political Parties Dialogue in November 2017 where foreign minister Wang Yi declared the goal was to "set an example to the world, prove that [China's] unique model, different from that of those traditional Western powers, can lead humankind, too."[88]

China's environmental health crisis is perhaps the most important litmus test for the resilience of the Chinese state. The environmental health crisis opens a window for political leaders to pursue desired policy change, possibly resulting in improved governance. As Dali Yang has observed, since the late 1980s, Chinese leaders have used crises and challenges to rebuild fiscal and institutional capabilities of the central state to "undertake painful but necessary economic and administrative restructuring" toward a more competent bureaucracy.[89] Invoking the old Chinese saying, "more disasters strengthen a nation" (*duo nan xing bang*), then-premier Wen Jiabao commented during the 2002–2003 SARS crisis that "a state and a nation can advance further every time it overcomes a difficulty."[90] If so, efforts to deal with the environmental health crisis may lead to institutional innovations that further the state's ability to handle similar changes in the future. Furthermore, the crisis, and the way it is handled, can be used to evaluate the state's crisis management skills and governance performance. If the leadership manages to undertake measures to reduce environmental health hazards in a significant and sustainable manner, that success would constitute powerful evidence of the political system's ability to adapt to emerging complex sociopolitical challenges.

On the other hand, if the government fails to address the crisis effectively – as seems likely given the magnitude of environmental health problems and the expanding gap between economic and social development – growing public frustration over a worsening environment may evolve into a larger sociopolitical crisis that threatens the very survival of China's political regime. If that happens, contemporary policy and institutional experiments aimed at tackling the environmental health crisis – such as modest governance reform measures – should be viewed only as a desperate ploy by an ultimately doomed regime whose "resilience" is nothing but a facade.

ANALYTICAL FRAMEWORK AND ORGANIZATION OF THIS BOOK

This book is interdisciplinary, drawing upon perspectives and findings from a combination of research on environment, public health, political economy, public policy, comparative politics, and international relations. When examining the environmental health problems and their impacts, I draw on (1) public health and environmental studies to demonstrate the health effects of environmental degradation; (2) economic models and analyses to gauge the costs of pollution attributed to the impact on health; and (3) concepts, approaches and models of comparative politics and international relations to discuss the impact of environmental health challenges on China's social-political stability and foreign policy. This is followed by an examination of government response to the environmental health challenges. Because of the focus on policy analysis, I will use both qualitative and quantitative data to investigate how the advent of the Xi Jinping era has affected agenda setting, policy formulation and enforcement, as well as policy outcomes.

Building on an analytical framework that places political leaders, bureaucratic officials, and social forces all in the same policy process, the study of the politics of China's ongoing environmental health crisis would be one of the few studies that addresses all key dimensions of contemporary Chinese politics, including elite politics (by discussing the evolving power politics of China's political elites); bureaucratic politics (by analyzing the interests, influences, and interactions of officials in different and often competing government agencies); central-local relations (by examining the extent to which interests and behaviors of subnational leaders are aligned with the central government leaders); state–society relations (by investigating the impact of social forces in policy making and implementation); and political development (by looking at the state capacity and prospect of China's democracy). Unlike the traditional bureaucratic politics model,[91] the analysis embeds bureaucratic capacity in the evolving political institutions in China and examines how the changing institutional context alters the opportunities and constraints faced by the policy actors, which in turn affect these actors' motives, expectations, and behavioral patterns in the policy process.

The rest of this book is organized into two parts.

The first part, beginning with Chapter 1, uses empirical data and scientific findings to demonstrate the health effects of environmental degradation, especially air, water, and soil pollution. The cancer village phenomenon epitomizes China's environmental health crisis. The analysis is unfolded in the context of China's rapid modernization and industrialization. Chapter 2 discusses the economic cost of environmental health problems, the crisis' impact on China's social-political stability, as well as the implications for China's foreign policy.

The second part focuses on the government's response by looking at policymaking and implementation and assessing the effectiveness of government response. Chapter 3 uses process tracing to examine the evolution of China's environmental health policy, including how domestic and international actors worked to elevate pollution control on the government agenda and the limitations and dilemmas in environmental health policymaking. Chapter 4 delves into the context and process of policy implementation, covering the roles of MEP and sub-national governments as well as the 2017 campaign toward meeting the air quality targets. Chapter 5 focuses on an assessment of policy effectiveness. In so doing, I examine the congruence between policy goals and outcomes and the follow-through on the steps and subgoals stipulated in the policy documents.

The book concludes with a summary on how the environmental health crisis is undermining China's rise, as well as implications for China's political development. I contend that the crisis and government response reveal a Chinese state whose political system is both resilient and fragile, and that the China model does not constitute a viable alternative to liberal democracy.

PART I

THE HEALTH AND NONHEALTH IMPACTS

CHAPTER 1

Health Effects of Environmental Degradation

GROWING UP IN A VILLAGE BY THE YANGTZE RIVER, I WAS accustomed to what many Chinese consider luxury goods today: blue skies, clean water, and a harvest season of golden paddy fields. It wasn't until I started middle school in a neighboring town three miles away that I was introduced to the idea that something might be amiss with my environment. On days when the tap water supply was interrupted, we had to fetch water from a river that ran through town to steam rice for lunch. Upriver from the town were several factories. I remember once asking a classmate whether the water was safe. "Yes," he replied, "the school chemistry teachers tested the water quality and they did not find anything harmful."

THE PATH TO AN ENVIRONMENTAL CRISIS

That was the early 1980s, a time when so-called township and village enterprises (TVEs) were springing up in China. Between 1978 and 1987, the number of TVEs increased more than tenfold, from 1.52 million to 17.5 million.[1] Since then, unleashing private enterprise and promoting unbridled economic growth have elevated China to the exalted status the founding fathers of the People's Republic only dreamed of when Mao Zedong declared in the 1950s that the nation would "surpass Britain and catch up with America" in output of iron, steel, and other major industrial products.

China's crude steel production tripled between 1980 and 2000 and then increased another fivefold by 2015.[2] In 2014, China churned out 823 million tons of steel, more than the rest of the world combined.[3] The steel industry is heavily concentrated in the area in and around Beijing–Tianjin–Hebei (BTH), China's equivalent to Germany's densely packed Ruhr region. With only 2 percent of China's landmass, the BTH region accounts for more than half of China's blast furnaces and contributes one quarter of the world's steel production.[4] In 2011, the fifty million tons of steel production capacity that the Hebei provincial government *failed* to report (to evade the central government's energy-saving and emission-reducing regulations) was greater than Germany's total output.[5]

Between 1981 and 2014, energy use in China increased by 275 percent.[6] During 2011–2013, China consumed more cement than the United States did in the entire twentieth century.[7] Overall, China consumes 60 percent of the cement, 49 percent of the steel, and 21 percent of the energy in the world.[8] Further, it utilizes 44 percent to 61 percent of the world's copper, coal, and aluminum.[9]

Rapid economic growth, urbanization, and the rise of a sizable middle class also fueled demand for motor vehicles. In 1990, car ownership for the civilian population was only 5.5 million. By the end of 2016, the number reached more than 194 million.[10] If all vehicles are included, the number rises to 290 million – compared to 269 million in the United States.[11] In 2017, more than 550,000 new vehicles took to the roads every week.[12]

China's economic accomplishments are mind-boggling. But so is the scale of environmental devastation they have left in their wake. China is now the world's biggest emitter of carbon dioxide (CO_2), sulfur dioxide (SO_2), and nitrogen oxide (NO_x). Wang Yuesi, a senior researcher at the Beijing-based Institute of Atmospheric Physics, estimated that nearly 90 percent of the smog in China was directly associated with human economic and social activities.[13] In Hebei province, the emissions of SO_2 and industrial dust in producing a ton of steel were 1.23 kg and 0.77 kg, respectively, in 2012. Both figures are three to six times the emission levels in industrialized Germany and Japan.[14] It comes as no surprise that the province, with a population of 74.7 million, claimed seven of the ten Chinese cities with the worst air quality in 2014.[15]

After four decades of unprecedented economic growth, it is also no surprise that hundreds of millions of Chinese, having been lifted out of poverty and into at least modest prosperity in little more than a generation, want something more: to live healthier lives in a cleaner environment. But calls for a shift to more environmentally sound industrial and energy structures frequently have been overshadowed by the perverse incentives embedded in China's political and economic institutions. Local government officials driven by career concerns still have strong incentives to promote growth in their localities; they often continue to welcome polluting technologies, which tend to be cheaper and contribute to output faster than clean technologies.[16]

In the village where I grew up, one of my childhood friends' fathers opened a family workshop in the early 1980s to manufacture wooden brushes. The by-product (wood dust) was dumped into the local river, and its decomposition caused a high concentration of nutrients in the river the villagers relied on for their water supply. Fellow villagers protested, and the factory was later relocated, but the practice of dumping wood dust continued until it became common to recycle it to make fiberboard.[17]

In rural areas, the abandonment of collectives and the return to household agriculture encouraged millions of farmers to spend more on chemical fertilizers and pesticides. When that led to a surplus of grain by the mid-1990s, the Chinese government encouraged a shift in production and food consumption. By 2017, China was producing 67 percent of the world's vegetable supply, 70 percent of the world's freshwater products, and about half of the world's hogs,[18] leading to even greater investment in fertilizers and pesticides. A 2010 report from China's Renmin University and Greenpeace said that Chinese farmers used 40 percent more fertilizer than crops needed.[19] From 1978 to 2011, the usage of chemical fertilizer increased by nearly sevenfold even though agricultural output increased by only 87 percent.[20] Pesticide use follows the same pattern.[21] Today, China still uses about 2.7 times as much fertilizer and twice as much pesticide per hectare as the world average.[22] China's agricultural authorities have indicated that the phaseout of the last ten highly toxic pesticides will not be completed until the end of 2022.[23]

A large amount of these fertilizers and pesticides has been discharged into water, further polluting rivers and lakes in the countryside. Indeed, by the late 1980s, when I left my hometown to start college life in

Shanghai, most of the local rivers and ponds had died: not only was the water no longer safe for drinking or swimming, but fish, shrimp, and river snails – important sources of animal protein amid a grain shortage in Mao's China – had disappeared.

The sheer size of China's economy and population only magnifies the environmental crisis triggered by rapid industrialization and the poor regulation of chemical disposal. According to the State Environmental Protection Agency (SEPA), in 2006, the banks of the Yangtze and Yellow Rivers were home to over half of the country's 21,000-plus chemical plants, which, together with power, paper, textile, and food production facilities, became a leading source of pollution in rivers, lakes, and groundwater in China.[24] Nationwide, over 70 percent of lakes and rivers in China were polluted and 78 percent of the water was not fit for human consumption.[25] Government officials admitted that in 2011 up to 40 percent of China's rivers were "seriously polluted" due to the sewage and waste discharged into them, and the water quality of 20 percent of the rivers was rated "too toxic even to come into contact with."[26] Incidents of chemical spilling and water contamination continue to be commonplace. Between 2007 and 2015, China reported 4,371 environmental emergencies, mostly caused by pollution.[27]

Air pollution has gotten much worse as well, driven by industrialization and urbanization. While health risks caused by ambient air pollutants can be traced back to the Mao era, the issue became much more serious and more widespread in the 1980s, especially in major Chinese cities.[28] A comparison with other countries revealed that as early as the mid-1980s, air quality (measured by the concentration of total suspended particulate or TSP) in major Chinese cities was already much worse than the peak pollution levels in Pittsburgh or Tokyo.[29] A report released in 2010 by the Ministry of Environmental Protection (MEP; the successor of SEPA) showed that about one-third of 113 cities failed to meet China's national air standards.[30] Nationwide, the average number of days with smog peaked in 2013 at 35.9 (the average number of smog days nationwide).[31] In three of the largest and most important urban metropolitan areas (BTH, the Yangtze River Delta, and the Pearl River Delta), there were more than 100 smog days annually.[32] According to a 2007 World Bank study, only 1 percent of China's urban residents breathed air at a pollution level deemed safe by European Union standards.[33] From April through July in 2014, 92 percent of the population was found to be exposed to more than 120 hours of air that were deemed "unhealthy" by the US Environmental Protection Agency (EPA).[34] A US-based nonprofit group, Berkeley Earth, estimated that 80 percent of Chinese citizens are regularly exposed to pollution levels much higher than those considered safe by the US EPA.[35]

A dominant contributor to the air pollution in China is PM2.5, which in some northern cities has reached 40 to 60 times the maximum level deemed safe by the World Health Organization (WHO). In 2014, the ten most-polluted Chinese cities all registered their daily average exposure at more than 100 micrograms per cubic meter ($\mu g/m^3$). According to American journalist James Fallows, that means even the worst-polluted American cities would rank among those with the best air quality in China. "No one now alive has experienced anything similar in North America or Europe," Fallows wrote, "except in the middle of a forest fire or a volcanic eruption."[36]

China has a soil pollution problem, too. Soil pollution is defined as "the presence of toxic chemicals (pollutants or contaminants) in soil, in high enough concentrations to pose a risk to human health and/or the ecosystem."[37] China not only has many contaminated areas in or near cities that once were used for industry ("brownfield sites") but also has large swaths of farmland contaminated by polluted air and water and/or an overuse of chemical fertilizers or pesticides. In the absence of enough fresh water to go around, farmers have used untreated wastewater and industrial effluent for irrigation. Indeed, only 10 percent of the sewage produced in rural areas is treated.[38] A national soil survey published by the MEP in March 2014 showed that nearly one-fifth of China's farmland was contaminated by chemical pollutants, with heavy metals, including cadmium, arsenic, nickel, lead, and mercury, being found in more than 82 percent of the contaminated land.[39] Cadmium, one of the most toxic elements in the environment, is the leading contributor to heavy metal contamination, affecting 13,300 hectares of farmland in eleven Chinese provinces.[40] The country has a long history – more than five decades – of exposure to heavy metal toxins.[41] The level of soil contamination in China is also much higher than in other comparable countries, such as Ukraine, which has 8 percent of its land contaminated.[42]

IMPLICATIONS FOR PUBLIC HEALTH

Official data reveal measurable advances in improving people's health standards over the past four decades. According to the State Council Information Office's 2017 white paper on China's health, in the period from 1981 to 2016, average life expectancy increased from 67.9 to 76.5 years old, while the infant mortality rate dropped from 34.7 deaths per thousand live births to 7.5 per thousand.[43] This is no small accomplishment for a country whose population is more than four times that of the United States, where the average life expectancy in 2015 was 78.7 and the infant mortality rate was 5.9, and whose average income, even taking into account differences in purchasing power, is roughly one-fourth that in the United States and Western Europe. By 2015, China had built the world's largest infectious disease surveillance and online emergency reporting system. Health insurance coverage has expanded significantly. Today, more than 95 percent of Chinese residents participate in some kind of health insurance, and out-of-pocket spending as a percentage of total health expenditure has dropped to 29 percent,[44] which is similar to Brazil – a standard bearer of universal health coverage among major developing countries.

Despite the remarkable progress, China is plagued by disease and ill-health. Data provided by the Ministry of Health (the predecessor of National Health and Family Planning Commission or NHFPC) suggested that the percentage of people who said they had been ill within the previous two weeks increased from 14 percent in 1993 to 24 percent in 2013.[45] Noncommunicable diseases (NCDs) now account for 87 percent of the disease burden in China, much higher than the global average (61 percent) and almost the same as the United States (88 percent).[46] Among NCDs, cardiovascular diseases (CVDs) are the biggest killer. About 21 percent of the population, or 290 million people, suffer from CVDs, which cause 3.5 million deaths annually and account for 41 percent of all deaths in China.[47]

The second-biggest killer is cancer, which accounts for 23 percent of all deaths in China. Forty years ago, cancer was barely known in rural

China. The first time I heard about it was in the early 1980s, when one of the villagers – who was in his fifties – was diagnosed with lung cancer. Hearing villagers talk about it as a terminal illness was spine-chilling for a teenager. Within one decade, cancer became a true epidemic in the country. Some of my loved ones were overwhelmed by this tide. At the end of May 1989, when I was in Shanghai watching the country inching toward the final showdown between the government and student protestors, my grandma – who largely raised me – died of cancer of the esophagus. Twenty-seven years later, my mother – who happened to be my first teacher at elementary school – took her last breath after fighting against lung cancer for nearly three years.

Today, while cancer incidence rates in China are still lower than the United States, they are rising rapidly, as are deaths from cancer. China had an estimated 4.3 million new cancer cases and 2.8 million deaths in 2015, compared to 900,000 cancer patients and 700,000 cancer-caused deaths in the early 1970s.[48] The number of annual crude cancer deaths per 100,000 more than doubled to 201 from 74 in the period from 1972 to 2015 (Figure 1.1). According to the Chinese National Cancer Center, 22 percent of global cancer incidences and 27 percent of global cancer deaths are found in China.[49] This rapid rise in the incidence of and death from cancer in China occurred at the same time that cancer incidences and deaths declined in many Western countries.[50]

Of all cancers, lung cancer is the most common in China. In 2015, China recorded 730,000 new cases of lung cancer, accounting for

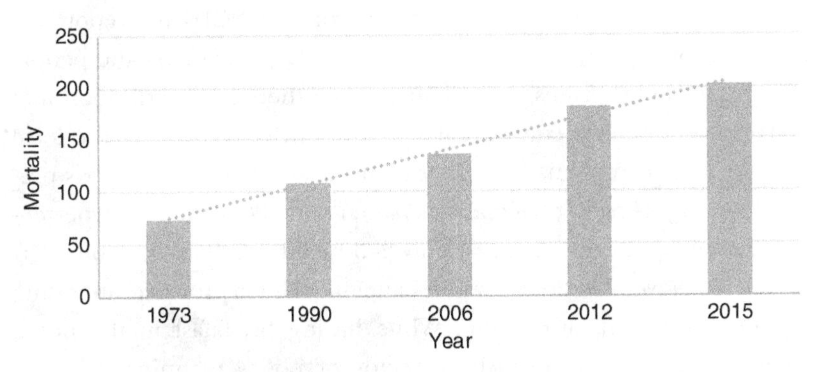

Figure 1.1 Rising cancer mortality in China, per 100,000 population (1973–2015)
Source: Author's database.

17 percent of total new cancer patients.[51] It is estimated that by 2025, China will have the world's largest population with lung cancer (one million).[52] The rising incidence rate of lung cancer coincides with drastic reductions in the incidence rates of stomach and cervical cancer (which are thought to be a result of improvements in public health conditions). Lung cancer is also the top killer of both men and women, in both urban and rural areas in China.[53] The number of lung cancer–related deaths in China increased by 465 percent between 1973 and 2013.[54] In 2017, 700,000 Chinese people – a number about the size of the entire population of Washington, DC – died of lung cancer, accounting for nearly 25 percent of all cancer deaths in China.[55]

The third leading cause of death is chronic obstructive pulmonary disease (COPD), an inflammatory lung disease that makes it increasingly difficult to breathe. About 8.6 percent of the adult population, or nearly 100 million people, suffer from this chronic lung disease in China.[56] COPD deaths account for 11 percent of all deaths in China, compared to around 5 to 6 percent in the United States.[57]

In addition to bodily ailments, more Chinese are suffering from mental health problems. From 2003 to 2008, recorded incidences of mental disorders jumped by over 50 percent. About 180 million Chinese – one in eight – suffer from some kind of mental illness.[58]

Unfortunately, neither the state nor the society is ready to cope with the growing burden of noncommunicable diseases. Despite the remarkable expansion in health insurance coverage, healthcare costs have increased steadily while the insurance benefit level remains low. The government has set up a surveillance network for NCDs and reportedly also has provided primary care management for the elderly and people suffering from hypertension or diabetes. But there is no indication that NCD prevention and control have received priority status at the local level. Primary care management of diabetes and hypertension was provided for only 24 percent of diabetes patients and 36 percent of hypertension patients.[59] In each county, only 2–3 health officials work on NCD issues.[60] In November 2018, I was in Langfang, Hebei province, attending an international health forum. While having breakfast in the hotel, I happened to sit next to the director of policy planning office of a provincial health department, who immediately became emotional when I brought up the issue of NCDs: "Please, help us let the central leaders know how serious the NCD problem is," he implored.

THE HEALTH IMPACTS OF POLLUTION: A PRIMER. To what extent does pollution contribute to these public health challenges? Until the 1850s, Western medical science on the etiology of disease was dominated by the miasmatic theory, which attributed the onset of diseases to a miasma (μίασμα, ancient Greek: "pollution"). In medieval Europe, the need to stay away from the noxious form of "bad air" led to the construction of lazaretto, a quarantine station holding and disinfecting humans and cargo, and the institution of cordon sanitaire, a guarded line preventing anyone from leaving a disease-ridden area. Later, Edwin Chadwick, a British social reformer who believed in the miasmatic origins of diseases, took up the question of clean water, sewage, and public sanitation and used quantitative methods to demonstrate direct linkage between poor living conditions and disease, which led to the Public Health Act of 1848. While the miasmatic theory became obsolete with the development of germ theory by Louis Pasteur and Robert Koch, environmental factors have continued to justify public health interventions worldwide.

China's "patriotic hygiene campaign," with its emphasis on sanitation and personal hygiene as well as safe and clean water, was driven primarily by the belief that a clean environment is crucial for people's health.[61] Public health experts agree that a person's immediate and concrete living conditions, including housing, water and sanitation, nutrition, food security and safety, and neighborhood conditions, can shape exposure, susceptibility, and resistance to illness, even death, at household and community levels.[62] In 2008, a WHO report further identified structural drivers (political and institutional context, social-economic policies and programs) and the resultant daily living conditions as areas of social determinants of health that needed to be addressed.[63] The international health agency estimates that living or working in an unhealthy environment caused 12.6 million deaths – nearly one in four of all global mortality – in 2012.[64]

While environment-related infectious diseases or social determinants of health are important concerns, the health effects of air, water, and soil pollution are becoming increasingly prominent. Thanks to improving sanitary conditions and better access to immunization and essential medicines, environment-related deaths from infectious diseases like diarrhea and malaria are in decline. As a result, NCDs now account for two-thirds of worldwide environment-related deaths, which are mostly attributable to pollutants from industrial emissions, vehicular exhausts, and toxic chemicals.[65]

Major environmental health disasters were well chronicled in the twentieth century. They include:

- The 1930 episode in the Meuse Valley, Belgium, where a thick smog containing sulfur and other pollutants caused up to 60 deaths and several thousand cases of pulmonary attacks;
- The 1948 smog in Donora, Pennsylvania, which killed 20 people and made 7,000 sick;
- The Great London Smog, which covered the British capital for five days in December 1952, killed 12,000, and hospitalized 150,000;
- The Bhopal disaster, a gas leak incident in December 1984 at the Union Carbide pesticide plant in India, which caused as many as 16,000 deaths;
- The Chernobyl accident, which occurred in April 1986 at a nuclear power plant in northern Ukraine and was responsible for an eventual death toll of up to 4,000 due to high level of radiation exposure.

In comparison to outbreak events like these, the harmful effects of chronic exposure to toxic chemicals are often not recognized and defined as a problem until they hit the headlines. In 1956, for example, Japan identified the first case of what became known as "Minamata disease," a chronic neurological disorder caused by mercury poisoning. The disease was responsible for 2,265 human cases and continued human deaths over more than 30 years.[66] Over the past two decades, scientists have documented the harmful effects of pollution on health[67] and described how exactly pollution affects health.[68] More recently, a Columbia University research team showed that local pollution levels were a better predictor of longevity in the United States than smoking, obesity, or socio-demographic features such as race or income.[69] Research also demonstrates the profound impacts of pollution on the health of children. According to the WHO, more than one-third of all childhood deaths can be attributed to the effects of a poor environment. This, together with rising numbers of birth defects and of childhood diseases, including autism, asthma, and obesity, raises serious questions not only about how toxic chemicals affect children's health[70] but also about whether or not pollution "threatens the continuing survival of human societies."[71] Global warming only exacerbates the health effects of pollution. Research suggests that synergies between toxic air pollutants and CO_2 can damage children's health by hampering cognitive and behavioral development and causing respiratory illnesses and other chronic diseases.[72] Growing evidence also shows that heat waves can interact with high levels of particulate matter (PM) and ozone and harm human health.[73]

Worldwide, one-sixth of all deaths are linked to pollution. Poor air quality – caused by both household (indoor) and ambient (outdoor) pollution – is identified by the WHO as "the world's largest single environmental health risk."[74] More than 90 percent of the world's population suffers toxic air, which is linked to seven million deaths annually. This led the WHO Director General, Tedros Adhanom Ghebreyesus, to call air pollution the "new tobacco."[75] Ambient air pollution alone contributed to 10 percent of global mortality, making it the fourth-highest global risk factor for death after diet, high blood pressure, and tobacco, and the leading environmental risk factor for disease.[76] A primary cause of air pollution is the burning of fossil fuels, which is also considered a major contributor to climate change, an issue associated with additional environmental health problems such as malnutrition and heat stress. That is why the WHO identified air pollution and climate change as among the top ten threats to global health in 2019.[77]

Of all the air pollutants that are a major public health concern – PM, carbon monoxide, ozone, nitrogen dioxide (NO_2), and SO_2 – PM2.5 has the greatest impact on human health. A recent cohort study of more than 4.5 million veterans in the United States reveals that 99 percent of the burden of death due to nonaccidental causes and 99 percent of the burden of death due to NCDs were associated with PM2.5 levels below the EPA guidelines (12 $\mu g/m^3$).[78] Based on cohort studies of half a million adults living in large cities in the United States, scientists concluded that for every 10 micrograms per cubic meter ($\mu g/m^3$) increase of PM2.5, overall mortality of cardiopulmonary diseases and lung cancer increased by 6 percent and 8 percent, respectively.[79] According to the 2015 Global Burden of Disease Study, exposure to PM2.5 was responsible for 4.2 million deaths and 103.1 million disability-adjusted life-years (DALYs), which account for 7.6 percent of total global deaths and 4.2 percent of global DALYs. To be specific, ambient PM2.5 pollution contributed to 17.1 percent of ischemic heart disease, 14.2 percent of cerebrovascular disease, 16.5 percent of lung cancer, 24.7 percent of lower respiratory infections, and 27.1 percent of COPD mortality.[80]

Ambient PM2.5 is also associated with increased risk of diabetes. According to a study published in the *Lancet Planetary Health*, long-term exposure to PM2.5 contributed to 14 percent of total incident diabetes globally in 2016.[81] If the risk of exposure to ozone is included, ambient air pollution causes an additional 254,000 deaths and a loss of 4.1 million DALYs from COPD in the same year.[82] In addition to the impact on the lungs and heart, breathing in certain airborne pollution particles may allow them to enter the human brain, causing neurodegenerative diseases such as Alzheimer's disease, the most common form of dementia.[83] Some scientists estimate that PM2.5 could account for 21 percent of all dementia cases.[84]

Research also supports that water pollution inflicts significant harm on human health. According to the US CDC, 80 percent of all diseases in the developing world are caused by consuming contaminated water.[85] In addition to acute waterborne diseases such as diarrhea, water pollution also puts people at risk of noncommunicable diseases, including cardiovascular diseases, cancer, and neurological disease.[86] When a baby drinks formula made with nitrate-rich water, for example, the body converts the nitrates into nitrites, which bind to the hemoglobin in the body to form methemoglobin, decreasing the ability of blood to carry oxygen. The lack of oxygen can cause "blue baby syndrome," a potentially fatal condition characterized by an overall skin color with a blue or purple tinge among infants. Besides nitrate contamination, arsenic as a source-water contaminant is linked to cancers of the liver, lung, bladder, and kidney.[87] Certain population groups are particularly vulnerable to water pollution. They include rural residents, who are more likely to die from cancer than their urban counterparts due to lack of access to treated water,[88] and pregnant women, who, if exposed to chemicals, will tend to have children with low birth weights.[89]

Compared to the relatively clear links between air or water pollution and human health, our understanding of the health impacts of soil contamination remains limited. The large numbers of factors influencing soil and human health, as well as the complex mixtures of the contaminants in the environment and bodies, pose many methodological challenges.[90] According to the WHO, chemicals found in soil with known health effects include mercury, lead, cadmium, arsenic, benzene, asbestos, fluoride, and hazardous pesticides. They cause health problems ranging from cancers and neurological damage to kidney disease and skeletal and bone diseases.[91] Lead poisoning, for example, can affect multiple body systems, including interference with the development of the central nervous system, harming the senses of touch and hearing, and causing poor coordination. Young children are most vulnerable to the toxic effects of lead, because their small body mass and still developing nervous systems make them more likely than adults to be exposed to and absorb lead.[92]

Overall, pollution is the largest environmental cause of morbidity and premature death in the world. In October 2017, the *Lancet* published an in-depth study showing that pollution-caused disease was responsible for an estimated nine million annual premature deaths in 2015 (16 percent of all global deaths), a number three times larger than mortality from AIDS, tuberculosis, and malaria combined. Of the nine million deaths, 92 percent were caused by air and water pollution.[93]

While China is certainly not alone in suffering from the health effects of environmental degradation, it stands out as one of the worst. According to the WHO, the environmental burden of disease accounts for 21 percent of the total disease burden in China.[94]

HEALTH IMPACTS OF AIR POLLUTION IN CHINA. Unlike many Western countries, where air pollution is limited by locality and duration,

China has found air pollution to be a nationwide and sustainable threat to health. The 2015 Global Burden of Disease Study found that each year 1.8 million people in China, or one-fifth of the global total, die from pollution, making China number two globally in total pollution-related deaths, behind only India.[95] This estimate may vary depending on what databases and methods are used in making the calculation. Mapping air pollution concentration and sources using data from 1,500 monitoring stations, two American scientists estimated that air pollution contributed to 1.6 million deaths per year in China.[96]

And unlike most Western economies, China's air pollution contains a complex mixture of pollutants, the distribution of which varies across regions and over time.[97] In addition to high concentrations of PM, China has high levels of SO_2, usually caused by coal-fired flue gasses, and NO_2, which primarily comes from vehicle exhaust. What Chinese call *wumai* ("haze") is a mix of the London fog-type particles and sulfates arising from fossil fuel combustion and Los Angeles smog-type photochemical pollution occurring in areas with high vehicle traffic and sunlight. Although the mechanism of haze formation in China remains unclear,[98] *wumai* consists of PM10, PM2.5, ozone, NO_2, SO_2, and carbon monoxide, all of which have health effects that range from exacerbated respiratory and cardiovascular conditions to a decline in lung functions.[99] The potentially higher health risk of pollution in China is indicated by a 2018 national report that records a higher lung cancer incidence rate among women in China than in Western countries (where more women smoke tobacco).[100] For this reason, existing environmental health models based on the Western experience may not be a good fit for analyzing the pollution-driven health impacts in China. As two Chinese scholars have pointed out: "Much of what we know about the marginal effects of pollution on health is derived from data reported in developed countries, where pollution levels are relatively low."[101] Research carried out by US scientists in 2017, for example, observed significant and positive correlation between PM2.5 concentration and depression and anxiety symptoms in the United States, but the mean annual concentration of PM2.5 in the study population was no more than 11.1 $\mu g/m^3$, much lower than China's PM2.5 concentration level.[102]

Of all the risk factors for pollution-related mortality in China, ambient PM2.5 is the number one killer, responsible for more than 1.1 million deaths (one quarter of total global attributable deaths) and 21.8 million DALYs (21 percent of attributable global DALYs).[103] In 2013, of the three million deaths in seventy-four of China's biggest cities, nearly one-third were associated with PM2.5. Of those who died from CVDs, lung cancer, and respiratory diseases, the PM2.5-related mortality accounted for 35 percent, 46 percent, and 16 percent, respectively.[104] At the provincial level, lung cancer deaths attributable to PM2.5 are the highest in Hebei, China's air pollution ground zero, and the lowest in Hainan and Tibet, which have the best air quality in China.[105]

The correlation between PM2.5 and incidences of cardiovascular and respiratory disease is corroborated by an analysis of monitoring data during 2013–2015 in 272 Chinese cities, the largest epidemiological study conducted in the developing world.[106] Data also suggest that the health impact of PM2.5 has accelerated since 2000. A survey carried out in central China's Hunan province, in the city of Changsha, found that the prevalence of childhood asthma was 1.6 percent in 1990 and 1.65 percent in 2000 but jumped to 3.79 percent in 2010.[107] In 2016, China already had about 25 million asthma patients. The number is estimated to reach 45 million by 2025, including 12 million urban child patients.[108]

Coal burning has been the leading contributor of deaths caused by ambient PM2.5 nationwide.[109] In 2013, PM2.5 from industrial, power plant, and domestic coal emissions led to an estimated 366,000 premature deaths, or 40 percent of the deaths attributed to ambient PM2.5 (see Figure 1.2 for breakdown). PM2.5 becomes a particular concern in the winter in northern China, when demand soars for the use of coal for heating. By the end of 2016, 83 percent of the square footage of heated buildings in northern China was heated by coal-fired boilers, many of them particularly inefficient, consuming 400 million tons of coal equivalent.[110]

Similar studies have examined the impact of air pollution on average life expectancy in China. In 2013, a paper published in the *Proceedings of the National Academy of Sciences* estimated that air pollution was associated with a reduction in life expectancy of about 5.5 years in northern China, where the government policy of providing heavily subsidized coal for indoor

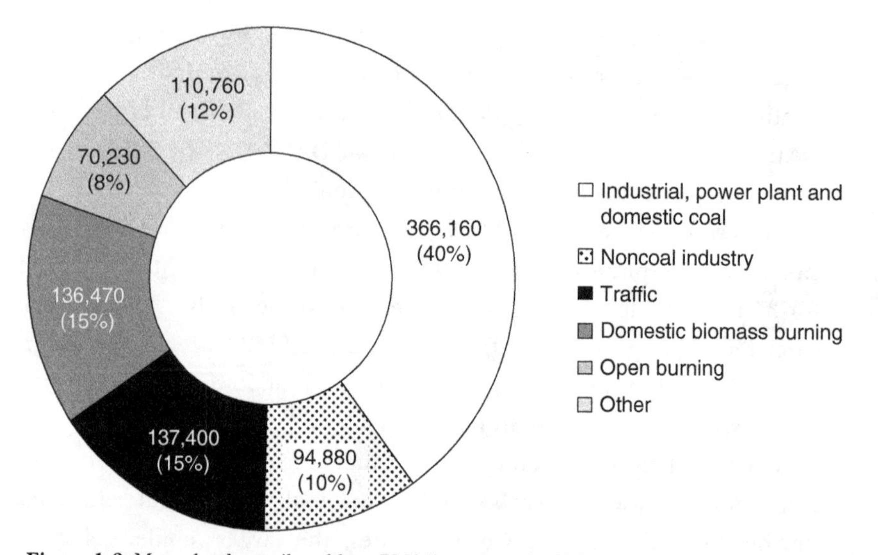

Figure 1.2 Mean deaths attributable to PM2.5 exposure in China, by source of emissions in 2013
Source: HEI Special Report 20: Burden of Disease Attributable to Coal-Burning and Other Major Sources in China www.healtheffects.org/system/files/GBD-MAPS-China-TableV-6-English.pdf.

heating during the winter resulted in sustained exposure to air pollution.[111] Coincidently, in 1999, then Chinese Premier Zhu Rongji half-jokingly told Beijing municipal government officials that "I would shorten my life at least five years by working here."[112] In September 2017, a new study covering a population much larger than the previous one showed that people in northern China live 3.1 fewer years than people in southern China. It also provided direct evidence of the health hazards of pollution particles, indicating that every additional 10 $\mu g/m^3$ of PM pollution reduces life expectancy by 0.6 years due to an increase in cardiorespiratory deaths.[113]

For years, public health experts have considered smoking to be the leading cause of premature deaths in China. Yet according to a study conducted jointly by Greenpeace and Peking University, the PM2.5-associated mortality rate was 90 in every 100,000 people, which was even higher than deaths from smoking (about 70 in every 100,000 people).[114] The findings corroborate another study by a team of scientists of Nanjing University, which suggests that health risks presented by ambient PM2.5 pollution are worse than those caused by smoking tobacco.[115] Indeed, fifteen types of carcinogens, including one of the world's strongest – benzopyrene – were found in China's PM2.5.[116] Recent studies also found that while the rising trend of smoking-caused lung cancer has been stabilized, lung cancer associated with the environment continues to increase.[117] A group of scientists analyzed historical records of aerosol particles and lung cancer incidence in the southern city of Guangzhou and found that a dramatic increase in the occurrence of air pollution from 1954 to 2006 was followed by a large increase in the incidence of lung cancer despite the drop in the overall smoking rate.[118] In other words, PM2.5 also may account for the rapidly rising incidence of lung cancer rates in China.

HEALTH IMPACTS OF WATER AND SOIL POLLUTION IN CHINA. Health problems caused by water pollution are a major contributor to disease in China. A survey carried out by a Chinese magazine in 2012 indicated that more than 70 percent of Chinese people feel threatened by water pollution.[119] Each year, diseases caused by water pollution make about 190 million people in China ill, killing an estimated 60,000.[120] In the Mao era, the focus of public health officials was on waterborne diseases such as schistosomiasis, a disease caused by parasitic worms. In 1956, Mao kicked off a campaign calling for "mobilizing the entire party and the entire society to eliminate schistosomiasis."[121] At that time, the toxic effects of industrial chemicals were barely on the radar. But beginning in the 1990s, a growing number of research studies examined the potential link between water pollution and cancer.[122] Yang Gonghuan, Deputy Director of the China CDC, led a research team to review the variation in trends in the causes of death in the Huai River basin over the past 30 years. Her team found that not only were the most seriously polluted areas the ones with the highest increase in digestive cancer deaths but also that there was a high level of overlap between the seriously polluted areas and areas with high cancer mortality.[123] The research findings were substantiated by a study that showed water pollution levels and distances from polluted water were highly correlated with mortality rates of esophageal cancer.[124] Using nationally representative data, another study examining the effects of industrial water pollution on individual health outcomes in China revealed that the degree of water pollution had a significant and independent effect on people's mental and physical health.[125]

The health impact of water pollution is further complicated by exposure to heavy metals.[126] While episodes of serious heavy metal contamination can be traced back to the 1970s and earlier,[127] they have become more frequent in the twenty-first century, with the ten most severe heavy metal contamination events occurring between 2005 and 2012.[128] In 1990, the average concentration of cadmium in Chinese soil was 0.097 mg/kg. Fourteen years later the concentration was equal to or exceeded the officially set environmental standard (0.3 mg/kg), indicating that a large amount of cadmium has polluted the soil.[129] The 2014 national survey reveals that 7 percent of soil in China is contaminated with cadmium.[130] Another study identified 50 incidences of mass lead poisoning (each with more than 19 cases) between 2004 and 2012, with 64 percent of them occurring after 2009.[131]

Heavy metals pose a health risk to all segments of the population in China, particularly children. A review of research published on children's blood lead levels (BLLs) found that during 2001–2007 nearly 24 percent of children's BLLs in China were higher than 100 μg/L (the amount of lead the US CDC identified as a level of concern).[132] Official Chinese media cited international public health experts to warn that if no effective measures were undertaken to address lead poisoning among children, the average IQ of Chinese would be lower than Americans by 5 percent.[133] A joint study conducted by Chinese and Swiss scientists noted that consumption of arsenic-laden groundwater could pose a risk to the health of nearly twenty million people across China, most of whom live in drought-stricken poor areas, such as Xinjiang, Inner Mongolia, and Shanxi.[134] The distinctively high risk posed by heavy metals in China is evidenced in research that found that New Yorkers born in Mainland China have higher blood levels of lead, cadmium, and mercury than both other New Yorkers of Asian heritage and all other New Yorkers.[135]

The consumption of contaminated crops constitutes a direct threat to food safety. As a group of Chinese scientists pointed out, "nowhere has that situation been more complex and challenging than in China, where a combination of pollution and increasing food safety risk have affected a large part of the population."[136] From 2003 to 2014, Chinese media have reported at least thirty-seven scandals involving fake or toxic food.[137] In the first three quarters of 2016 alone, 500,000 instances of illegal food safety violations (e.g., sale of contaminated food products) were uncovered by regulators in China.[138]

A major source of food poisoning is chemical contamination, which can be caused by man-made factors, including industrial pollution, traffic emissions, and agricultural activities such as the use of pesticides and chemical fertilizers.[139] Widespread coal burning, for example, releases lead, mercury, and arsenic that contaminate soil and can penetrate the food chain by entering animal feed and animal meat. These heavy metals pose one of the greatest risks to food safety and health in China, because they tend to be more mobile and can mimic the functions of nutritive minerals and thus be readily taken up by crop plants. According to an MEP document, 36,000 hectares of farmland contain "excessive" levels of heavy metals, leading to the contamination of twelve million tons of grains each year.[140] In eastern China's Anhui province, heavy metals in cultivated topsoil and grain were found to be correlated with human liver, lung, and gastric cancer.[141] Cadmium, in particular, can accumulate in the human body if ingested, causing joint pain, bone disease, and even cancer. The concentration and health risks of heavy metals, though, vary across cropping systems, with rice consumption being the most important single health risk.[142]

Because rice is a major staple food in China, it is particularly worrisome that it has become a major source for exposure to heavy metals. In 2002, in one of the few nationwide food safety tests, the Ministry of Agriculture (MOA) found that 28.4 percent of rice samples were laced with levels of lead that exceeded national safety standards, and 10.3 percent of them had excess cadmium.[143] Based on a more recent study, researchers concluded that contaminated rice is the leading contributor to cadmium exposure in China, responsible for nearly 56 percent of the total cadmium intake for the general Chinese population.[144]

The problem worsened as China modernized. In 2006, a group of China CDC scientists revisited Dayu County in Jiangxi province. They were astounded to find that the cadmium concentration in rice increased by 30 percent over the previous 19 years and had reached a level almost the same as that in the Fuchu area in Japan, where mass cadmium poisoning caused the outbreak of "itai-itai" disease, one of Japan's four most prominent pollution-caused disasters.[145] From 1990 to 2015 the mean dietary cadmium exposure of the general Chinese population more than doubled, from 13.8 μg per day to 30.6 μg per day.[146] In May 2013, food safety officials in Guangzhou announced that eight of the eighteen samples of rice and rice products from Hunan, China's largest rice-producing province, were heavily laced with cadmium.[147] Reports of cadmium rice in Jiangxi and Hunan immediately prompted consumers to switch to rice produced in northeast China, or even Southeast Asia.[148]

CANCER VILLAGES

For many Chinese, the price they are paying as a society for air, water, and soil pollution is most evident in the phenomenon of "cancer villages," small communities where cancer rates are far above the national average. One such village, Qingxia Community, is located in the industrial city of Zhuzhou, in Hunan province. Surrounded by dozens of highly polluting factories, at one point 10 percent of Qingxia's residents suffered from cancer, well above the national average of 1 percent.[149]

Chinese journalists began to use the term *Aizheng Cun*, or "cancer villages," as early as 1998. The government did not admit their existence until 2013.[150] In May 2009, a Chinese journalist, Deng Fei, unearthed all related reports and published a "Cancer Village Map," locating more than 100 cancer villages.[151] The same year, Sun Yuefei, a graduate of Central China Normal University in Wuhan, identified more than 247 such villages in 27 provinces in China, with the situation in Jiangsu and Henan being particularly severe.[152] Combining both official reports and online sources, Professor Lee Liu at the University of Central Missouri counted a total of 459 cancer villages in 2010 spreading across every province except Qinghai and Tibet.[153] Using the US CDC's definition of cancer clusters, two Chinese scholars reclassified the reported cases and identified a total of 351 "cancer villages" by the end of 2011.[154]

While "cancer villages" are similar to cancer clusters found in other countries, they differ from their international counterparts in terms of the origin, geographic distribution, and health effects. The first cancer village was identified in 1954 in southwest China's Yunnan province,[155] but they didn't really mushroom until the end of the 1980s. From 2000 to 2009, 186 new cancer villages were identified, accounting for 53 percent of all cancer villages.[156] Considering that it takes an average of 20 years for the pollutants to accumulate in the body to develop cancer, some Chinese scholars suspect that the rapid proliferation of cancer villages in the first decade of the twenty-first century was the lagged effect of pollution produced by the boom in TVEs in the 1980s.[157]

The theory that industrial pollution is the primary cause of cancer villages is supported by the dominance of chemical carcinogenic agents in these villages (e.g., nitrosamines, pyrrolizidine alkaloids, polycyclic hydrocarbons, and heavy metals), which were found in over 95 percent of the cancer villages in China, compared to the 2 percent of cancer villages caused by viral carcinogenic agents (e.g., HBV) and the 1 percent by physical carcinogenic agents (e.g., radiation, asbestos, fibers, and PM).[158] The theory also explains why a majority of the cancer villages (51 percent) are located in eastern China (where TVEs were concentrated and highly developed), compared to 36 percent in central China and 13 percent in western China. Cancer villages in China also tend to cluster along major rivers and their branches. Nearly 60 percent of them are located less than 3 km from a major river or its branch, and 81 percent of them are located less than 5 km from a major river or its branch.[159] This led some Chinese researchers to conclude that cancer villages in China are river basin–wide phenomena, in contrast to the more localized cancer clusters found in other countries.[160] One study suggests that a decline in water quality by a single grade increases the digestive cancer death rate by nearly 10 percent.[161]

Beginning in the 1990s, as most TVEs changed ownership or closed down, many local governments turned to large-scale external investment, often in polluting industries like chemicals, paper, and resource extraction and processing, to promote economic growth in their jurisdictions. Their large size and high pollution levels further exacerbated local environmental degradation.[162] Studies of China's industrial relocation in period 2004 to 2008 and period 2009 to 2013 also show that polluting industries have become more regionally concentrated and have increased their presence in the inland regions of China.[163] That suggests China will suffer from a growing number of cancer villages in central and western China in the next decade or two.

Villagers have responded in a variety of ways to the outbreaks. Some take a stoic attitude toward their suffering, falling ill and dying with cancer without resorting to expensive forms of treatment.[164] Others rely on passive preventive strategies like closing windows at night and shouldering the medical bills on their own. But others seek redress for harm by collecting evidence and statistics, pursuing compensation from polluting factories, petitioning the local governments for action, or seeking help from the press or NGOs. The patterns and outcomes of villagers' actions vary depending on how serious the pollution-caused damages are, whether villagers have family members working in the factories, how committed activists and leaders are to their protests and petitions, and the level of government accommodation, perceived or real.

Fieldwork suggests that the protestors and petitioners may not always be persistent in pursuing their causes: while residents in one cancer village were very assertive in constructing the image of a cancer village, those in the neighboring cancer village, after their demands were met, began to "destigmatize" in recognition of the "side effects" (e.g., pressures from local governments; difficulties finding spouses among young people).[165] Fear of government revenge often prevents persons in the know from speaking out. My sister-in-law once mentioned that in the village she grew up, six people died of cancer after working in a polycrystalline silicon factory. She immediately shut her mouth when I tried to seek more information.

The response from the polluting factories and government actors also varies. In some localities, polluting factories were closed down or relocated, and villagers were at least partially compensated for their losses. More often, villagers' protests and petitions were stonewalled by collusion between factories and government officials pursuing economic development, which swells local coffers and helps advance the careers of local officials.[166]

Overcoming local resistance often requires public attention. As Benjamin Van Rooij has noted, NGOs and media can serve as important intermediaries for citizen action against pollution.[167] Press reports can be particularly effective in spurring central leaders to exert pressure on local governments to take action. Indeed, the Five-Year Study of Water Pollution and Cancer in the Huai River Valley was launched after Premier Wen Jiabao saw media reports about cancer villages in the region.[168]

In confronting the polluting firms, one of the most difficult challenges the villagers and their outside supporters face is establishing a causal linkage between pollution and its carcinogenic effects, which is considered sine qua non for building a valid case. Establishing the linkage is different from assessing the health effects in an environmental emergency, where the culprit is relatively easy to identify.[169] True, the carcinogenic effect of particular chemicals is well understood by scientists and public health experts. And even without detailed information about specific pollutants, researchers can use regular environmental monitoring data, as they did in the Huai River study, to establish robust statistical correlation between environmental quality and its health effects. Still, given the scale of polluting industries in many areas, it is extremely hard to determine the contribution of a particular source of pollution to a particular cancer in a given location in China.[170] In 2010, under the instruction of then executive vice-premier Li Keqiang, MEP led a three-year study on the cancer villages, but the study only identified a very vague relationship between pollution and cancer.[171]

My hometown – a county-level city with an area of 332 square kilometers and a population of 300,000 – has one of the highest upper gastrointestinal cancer incidence rates in China. Since the 1970s, many studies have been conducted in order to identify the causes of such high cancer rates.[172] Even though these studies support the health effect of broadly defined environmental factors, they have been unable to establish a causal relationship or isolate a specific pollutant in explaining the high cancer rates.

If proving causality within a county is difficult, it proves even more challenging to do so at the village or individual level. Epidemiologists conducting statistical analysis – in order to isolate a particular environmental factor – have to take into account numerous, often confounding factors such as genetics, socioeconomic status, nutrition intake, and individual habits.[173] While this issue is not unique in China, rapid industrialization and social change have made this endeavor even more daunting.[174]

Without government cooperation, collecting evidence for an environmental health case is expensive (even the most basic tests can cost tens of thousands of dollars[175]) and, given the amount of pollution villagers are often exposed to on a daily basis, it becomes extremely difficult to isolate the carcinogenic effects of single pollutants or specific factories. In 2008, a fish farmer from one cancer village filed a court case against two local steel factories, which he believed dumped toxic waste that caused his two daughters to develop leukemia. The local court ruled against him, saying that he had not provided sufficient evidence to establish such a link.[176] Local government officials almost always side with well-connected businesses in dealing with the villagers, and local courts (whose decisions are subject to interference from local governments) regularly rule against the villagers in any lawsuits. As a result, the villagers who stand up to the pollution rarely prevail.

ADDITIONAL ENVIRONMENT–HEALTH LINKAGES

TRASH AND PUBLIC HEALTH. Garbage has become another major contributor to China's environmental health crisis. As early as 2004, China surpassed the United States to become the world's largest municipal solid waste (MSW) generator.[177] By 2017, China had amassed eight billion tons of trash, much of which is in landfills in the countryside.[178] Ten years ago, I travelled with a group of Chinese scholars to the outskirts of Beijing, where we saw an exposed garbage mountain sending its odors far and wide and lofting flurries of paper and plastic waste into the wind. At that time, I was not aware that it was just one of some 500 trash sites that encircle Beijing.[179] Today, such mountainous trash landfills have mushroomed across China and a large number of Chinese cities have come "under siege" from a deluge of trash that grows larger every year, contributing to pollution that, according to a Chinese journalist, is "more widespread, more serious, more profound, and more prolonged than smog."[180] A 2013 survey by the Ministry of Housing and Urban-Rural Development (MOHURD) showed that more than one-third of Chinese cities are encircled by garbage, and one-fourth of the cities have difficulties finding new sites to dump trash.[181] In 2018, a government newspaper reported that two-thirds of the 600 large and medium-sized cities in China were besieged by waste.[182] Unsurprisingly, there was a spike in news reports about illegal cross-region garbage dumping.

The problem is probably even worse in rural areas, where 40,000 townships – including nearly 600,000 villages – produce 280 million tons of household trash each year, most decomposing by natural means.[183] In 2010, the China City Environmental Health Association estimated that China produced 1 billion tons of garbage annually, including 400 million tons of household trash and 500 million tons of construction debris.[184] In 2015, only 180 million tons of garbage was properly disposed, and only two-thirds of that ended up in landfills.[185] Often illegally dumped in unofficial landfills and heaps, much of China's trash leaks pollutants into the ground, water, and air and gives off a nauseating stench.

These astonishing numbers do not include eight million tons of electronic waste (e.g., used TV sets, computers, and cellphones) that are smuggled into China each year. *China Business News* reported that China was the final destination of about 70 percent of the world's annual e-waste of 500 million tons.[186] Disposal of toxic e-waste has become a booming business in some Chinese localities. Guiyu, a small town in Guangdong province, may well be the largest e-waste site in China. *Caijing* magazine reported that 21 of the 27 villages in the town have been involved in the disassembly of e-waste and plastic recycling, hiring 100,000 of 160,000 of the local residents and receiving 1.55 million tons of e-waste and plastics each year. According to the former head of the township, each year Guiyu extracted no less than 15 tons of gold from e-waste, accounting for 5 percent of China's gold output.[187] To hold down costs, the workers relied on the most primitive means of extracting precious metals: circuit boards and microchips were either burned individually over open fires or soaked in highly corrosive and dangerous acid baths. These methods expose workers and local residents to various environmental hazards. While the smelting process discharges large amounts of toxic gases into the air, the acid baths often leached into local streams without treatment, contributing to water and soil pollution.

A joint study conducted by the Chinese Academy of Sciences and Sun Yat-sen University in Guangzhou concluded that the heavy metal concentration in Guiyu soil was 100 to 1000 times higher than comparable regions.[188] Greenpeace Research Laboratories found that the water was so acidic that it is "powerful enough to disintegrate a penny after a few hours."[189]

Garbage and improper waste disposal are suspected contributors to cancer and other health problems in China. In one village near a landfill in a Beijing suburb, nearly 20 people died of cancer in 2016, and a growing number of the villagers have been diagnosed with gallstones and kidney stones.[190] A study carried out by Professor Huo Xia of Shantou University Medical College found that 82 percent of Guiyu children had BLLs over 100 μg/L.[191]

As the amount of household and industrial waste continues to grow, and as new sites for landfills become difficult to find, the authorities are building more incinerators, which cause additional concerns about pollution and health. According to a report released by two Chinese environmental NGOs, only 65 of the 160 waste-to-energy (WTE) incinerators investigated by researchers in 2014 disclosed pollution emission data, and 45 of them were unable to meet China's new national standard.[192] Dioxin emissions from these WTE plants are of particular concern: dioxins are known to weaken the immune system, cause reproductive and developmental problems, disrupt hormones, and lead to cancer.[193] A group of Chinese scientists measured dioxin emissions in 19 WTE incinerators in China and found that three of them do not meet national standards and 16 do not meet EU standards.[194] Cancer villages now can be found near many waste incineration power plants in China.[195] The health effects of dioxins and other pollutants explain why there is fierce public opposition to the construction of waste incineration plants near residential areas.

ENVIRONMENT AND MENTAL HEALTH. The relationship between pollution and mental health has received a great deal of scholarly attention over the past decade. Scientists have found significant correlation between the levels of air pollutants and mental problems such as depression, anxiety, and suicide attempts.[196] An online survey found that during haze days, 45 percent of respondents felt dreadful, 23 percent of them were anxious and restless, and 82 percent had a low mood.[197] Poor air quality reduces short-term hedonic happiness and raises the rate of depressive symptoms in China. Increases in the Air Pollution Index (API), for example, account for nearly one quarter of the actual decline in happiness from 2007 to 2014.[198] Across all major air pollutants (PM2.5, PM10, NO_2, O_3, SO_2), PM2.5 is found to have the strongest effect on happiness and depressive symptoms.[199] From 2003 to 2012, Chinese scientists identified significant correlation between the quarterly rise in air pollutants and the increase in suicides.[200] In addition to the link between pollution and suicide rates, air pollution is also connected to non-suicidal self-injury among Chinese adolescent students.[201] High pollution levels can be considered an important contributor to overall malaise in China, where more than 180 million people – one in eight – suffer from some kind of psychiatric disorder.[202]

POLLUTION AND FERTILITY. Increased environmental pollution is suspected as a cause for the decline in the quality of sperm among Chinese men. A 2012 study found that only one-third of the semen at Shanghai's main sperm bank met WHO standards, and that pollution was considered one of the major causes.[203] In 2015, fewer than one-fifth of young men who donated sperm in Hunan province had sufficiently healthy semen to qualify as donors, down sharply from 2001, when more than half qualified.[204] The decline in sperm quality is considered a prime reason for a significant increase in the infertility rate for all couples of childbearing age, from 3 percent in 1990 to 12.5–15 percent in 2018.[205] Despite China's 2016 policy change allowing couples to have up to two children, instead of one, two Chinese demographers estimated that China's total number of births dropped below fifteen million in 2018, the third-lowest total since the founding of the PRC in 1949, exceeded only by 1960 and 1961, the years of extensive famine in China.[206]

"To anyone who has been following the last 25 years of research on human sperm count decline," John Peterson Myers, a leading environmental health scientist, wrote in an email exchange, "the pattern reported from China was not surprising because of the ubiquity of exposure to endocrine disrupting chemicals in China."

POLLUTION AND ANTIBIOTIC RESISTANCE. As the world's leading producer and consumer of antibiotics, China has faced huge problems of antibiotic abuse and overuse. Until recently, roughly 70 percent of all pharmaceuticals manufactured in China were antibiotics (compared with 30 percent in the West), and 70 percent of inpatients received antibiotics from the medical practitioners at Chinese hospitals (against the WHO-recommended maximum of 30 percent).[207] This does not include the ubiquitous use of antibiotics for poultry and livestock in China.

As a result, excessive amounts of antibiotics are present in China's rivers, where they can enter the human food chain and threaten to cause widespread resistance across drugs. According to a report published by the authoritative *Chinese Science Bulletin*, sixty-eight kinds of antibiotics have been detected in China's surface water, and the concentration of antibiotics in major China's rivers is much higher than that in more developed countries.[208] Overuse of antibiotics also has expanded antibiotics resistance reservoirs from water to soil. A study of samples from paddy soils from southern China identified sixteen types of antibiotic-resistance genes (ARGs), and in all samples, multidrug resistance genes were found to be the most dominant type (38–47.5 percent).[209] Air pollution only adds to the problem. In 2016, international scientists conducting analysis of DNA samples taken from humans, animals, and environments worldwide found that smog in Beijing harbored the largest number and types of ARGs, including several that can be resistant to carbapenems, considered the last-resort antibiotics for treating bacterial infections.[210]

POLLUTION FIRST, PROTECTION LATER?

In January 2018, the Council on Foreign Relations organized a roundtable meeting on China's environmental health. At the meeting, Jeffrey Sachs, a world-renowned development expert, raised the question: Is China's case unique?

In the 1970s, Chinese leaders considered environmental pollution an "incurable disease" unique to the decadent, capitalist West. Today, few would deny that China is beset by a plague comparable to what the United States, Europe, and Japan faced as they industrialized. But Chinese policymakers often take consolation in the idea that China is following a similar path of "pollution first and protection later." For many Chinese, pollution and its health consequences are the necessary evil released from the bottle of development. The implication is that China – like its Western counterparts – eventually will manage to place pollution and its health consequences under control.

What they have not considered is the monumental task China is now facing. Pollution in China is not a localized problem but a grave challenge found almost everywhere. Moreover, as we will see in the following chapters of this book, the causes and consequences of pollution are often more complex and multifaceted than those in Western countries, and the obstacles to overcoming them are even more daunting.

As Qu Geping, the first administrator of SEPA, observed:

> Worldwide, no other country faces such serious environmental pollution as China. In addition to air pollution, there are problems of water pollution, soil pollution, toxic chemical pollution, and so on. The problems are just too many, and too big. Ordinary Chinese's minimum and most basic health and safety are in danger.[211]

CHAPTER 2

Economic, Sociopolitical, and Foreign Policy Impacts

When visiting Berlin, a stop at the DDR museum provides a glimpse into ordinary life under Communist rule in East Germany. "One of the most interesting aspects was finding out how central environmental consciousness was to the rise and fall of socialism in [East] Germany," one visitor commented.[1] By the 1970s, East Germany, whose initials for its formal name in German were DDR, had become the most developed economy in the Soviet bloc. Yet that was also the time that some areas of the country began to experience severe environmental degradation, causing the levels of heart and respiratory diseases, as well as various types of cancer, to rise rapidly.[2]

Keenly aware of the threat an environmental health crisis could pose to its political legitimacy, the DDR regime chose to make such data a state secret. The cover-up, however, inspired further grassroots resistance, which coalesced into an environmental movement that called attention to the dangers of pollution and criticized the government's approach to the problem. Despite a government crackdown, the movement gained momentum by forming networks with other civil society groups and making connections with the West, all of which helped fuel the collapse of the Communist state.[3]

China differs from the DDR in many critical aspects, but it does resemble its former socialist ally in one important way: its heavy reliance on coal. Like East Germany, where coal once supported more than 80 percent of the energy needs, China has a coal-dominant energy mix, with coal at one point making up about 70 percent of its total energy consumption. And just like the DDR, China has been caught up in a public groundswell of environmental protests and concerns over the degradation of its air, land, and water.[4]

Does that mean China will fall into the same trap as East Germany did? Not necessarily, but Qu Geping, dubbed "the father of China's environmental protection," would not rule out such a possibility: "I think without abating pollution and protecting the environment, the party will be destroyed and the nation will be subjugated."[5] In the remainder of this chapter, I will explain why environmental health issues carry such profound implications for the future of China and how they threaten to severely weaken the nation's economic growth, undermine its sociopolitical stability, and complicate China's foreign relations.

ENVIRONMENTAL HEALTH AND THE NATIONAL ECONOMY

Economists long have observed the positive and significant effect of good health on economic growth.[6] According to the WHO Commission on Macroeconomics and Health, improved health has its most important economic effects on human capital through education, on-the-job training, physical and cognitive development. It also affects enterprise capital by enhancing teamwork, workforce organization, and an ability to attract labor and capital.[7] And the contrary is true as well: the burden of disease stands as a stark barrier to economic growth, due to higher medical costs, income lost through absences due to illnesses, reduced savings, and disposable income for households and increased government spending on welfare programs. Diseases also have long-term economic effects through their impacts on labor supply, human capital, and labor productivity.

Extending the health–wealth linkage to the environmental arena, economists have examined the relationship between pollution, health, and economic development. The Lancet Commission on pollution and health cited research demonstrating how improvements in air quality can lead to significant economic gains. Since the passage of the Clean Air Act in 1970, for example, every dollar the United States invested in air pollution control is estimated to have yielded $30 in economic benefits. Similarly, the removal of lead from gasoline in the United States, which led to increased cognitive capacity among children born after 1980, has translated into an aggregate benefit of over $6 trillion to date.[8]

In the meantime, a large number of studies have demonstrated the negative impact of pollution on labor supply, productivity,[9] and academic performance.[10] For example, research has suggested that a one-unit increase in lead exposure increases the probability of a child's suspension from school by 6.4–9.3 percent.[11] Air pollution not only increases the risk of dementia and other forms of cognitive decline but also raises the likelihood of poor behavior and decision-making.[12] Given China's extreme air pollution, the economic effects of high concentrations of PM2.5 have drawn particular attention from researchers.[13] Focusing on PM2.5's impact on healthcare spending, one study concludes that an increase of 10 micrograms per cubic meter ($\mu g/m^3$) in PM2.5 concentration results in a 0.31 percent increase in the total number of healthcare transactions in the short term and a 2.74 percent increase in the longer term. The long-term impact implies that a reduction of 10 $\mu g/m^3$ in daily PM2.5 concentration could result in annual savings of at least $11 billion in total health spending in China.[14]

Scholars have also examined the impact of air pollution on cognitive development, human capital formation, and labor productivity in China. They found that contemporaneous and cumulative exposure to air pollution significantly lowers both the verbal and math test scores given to people aged 10 and above.[15] They also identified a significant, but non-linear, relationship between ambient PM2.5 concentration and labor productivity. In 2007, PM2.5-caused diseases negatively affected productivity of about 72 million workers – 10 percent of the total labor force – in 30 Chinese provinces, costing the economy an estimated $44 billion, or about 1.1 percent of the GDP.[16]

Large drops in worker output – equivalent to 14 percent of mean output – were documented when PM2.5 levels rose from 10 to 210 $\mu g/m^3$, but the drop in output leveled off with further increases in PM concentrations. The research suggests that if the distribution of hourly PM2.5 were reduced to 25 $\mu g/m^3$ (the WHO guidelines for 24-hour mean exposure), labor productivity across 190 cities in China could increase by an average of 4 percent annually.[17] Driven largely by health impacts and loss of labor productivity, the annual cost of pollution in China between 2000 and 2010 was estimated by the World Bank and China's Development Research Center of the State Council to be close to 10 percent of GDP, with air pollution accounting for 6.5 percent, water pollution 2.1 percent, and soil pollution 1.1 percent.[18] This ratio is several times higher than in South Korea and Japan, and much higher than in the United States.[19]

Measuring the overall economic effects of environmental health in China is challenging, not only because of the lack of information about the costs of disease but also due to the lack of consensus on modeling the economic costs of environmental health problems. The so-called adjusted human capital (AHC) approach is the official method used to value mortality risks in China. Relying on AHC calculations, a World Bank study found that the economic burden of premature mortality and morbidity associated with air pollution alone was 157.3 billion *yuan* in 2003, or approximately 1.2 percent of GDP.[20] The actual costs of pollution are certainly larger than the estimate arrived at in the study, which did not take into account the costs of water and soil pollution.

One shortcoming of the study's method is that it is based solely on the effects of pollution on GDP and thus fails to include intangible losses, such as the negative effects of pollution on societal health, family disruption that follows the premature death of beloved ones, and destruction of ecosystems and loss of key species that are crucial to sustaining life on earth.[21] A think tank within the Ministry of Environmental Protection (MEP) estimated that if the lost productivity due to health issues, crop degradation, and losses from pollution-related accidents were included, total costs arising from pollution amounted to at least 512 billion *yuan*, or 3.1 percent of GDP, in 2004.[22]

With the rapid improvement of material living standards in China, the AHC approach also fails to take into account that Chinese people increasingly are valuing things beyond basic earnings, such as leisure, good health, consumerism, and life itself. When I was doing fieldwork in my hometown for this book, one older lady told me, "Now that the life is getting better, we don't want to die." Surveys conducted by Chinese economists in Shanghai and Chongqing in 2005 and 2006 found that the value of a statistical life (VSL) – an estimate of how much people are willing to pay to reduce mortality risk – was approximately one million *yuan* ($122,000), which was much greater than what is implied by the AHC approach (approximately 280,000 *yuan* or $34,200). More recent studies also suggest that Chinese residents are willing to pay on average

3.8 percent of their annual household income to reduce PM2.5 concentration by 1 $\mu g/m^3$, and families with children under the age of six are willing to pay 5.9 percent of their annual income (an average of $120) for a 1 $\mu g/m^3$ reduction per year compared to 3.3 percent ($70) for families without children below the age of six.[23]

Applying the VSL approach to China, the health costs associated with ambient air pollution go up to about 520 billion *yuan* per year, or about 3.8 percent of 2003 GDP. If combined with the health costs associated with water pollution, the approach produced a price tag of about 586 billion *yuan*, or 4.3 percent of GDP.[24] The health costs accounted for three quarters of the total cost of air and water pollution in China in 2003.[25] The actual costs could be higher because the World Bank study ignored morbidity associated with NCDs such as cancer and did not consider the costs associated with soil pollution and food safety.

In a more recent study, the World Bank and the Institute for Health Metrics and Evaluation (IHME) adopted a similar approach to estimate the economic costs of indoor and outdoor pollution worldwide.[26] The study found that the economic costs of air pollution have increased significantly over time in China. From 1990 to 2013, the annual growth of total welfare losses due to premature mortality caused by exposure to air pollution was 10.9 percent, highest among all other countries except Equatorial Guinea (13.8 percent). In 2013, air pollution in China was responsible for $1.59 trillion in welfare losses or 31 percent of the global total.[27] Put differently, by this estimate, China lost about 10 percent of its GDP to air pollution, the largest share among all the countries covered by the study. If we also include in the scope of the analysis the incremental costs incurred from nonfatal illnesses, the total costs could reach up to $2.83 trillion, or 18 percent of China's GDP.[28] Even if welfare losses due to premature mortality caused by water and soil pollution are discounted, China is still among the economies most adversely affected by pollution.

Findings on rising national costs related to pollution are consistent with analysis of subnational data. Research found that health costs related to Beijing smog, including both the direct cost of healthcare spending and the indirect cost of income loss due to absenteeism, increased by 140 percent between 2003 and 2013.[29]

And even after China vowed to tackle its environmental problems, there is evidence that it is losing ground rather than gaining it. Dreadful air pollution is becoming the top challenge for multinational corporations trying to recruit and retain foreign workers in China. Customer data from UniGroup Relocation, a company that helps relocate multinational employees, show twice as many expats moved out of China than arrived in 2014.[30] This trend is confirmed by a survey conducted by the American Chamber of Commerce in China, which revealed that between 2010 and 2015 the share of respondents reporting difficulties in recruiting senior executives to work in China due to air quality issues increased from 24 percent to 52 percent.[31] For 3 consecutive years between 2017 and 2019, poor air quality has been found to be the number one challenge to foreign companies' ability to recruit and retain talent to work in China.[32] Meanwhile, the percentage of executives agreeing that food and water safety concerns are a significant challenge increased from 28 percent to 36 percent.[33] Ding Xueliang, a Hong Kong-based China scholar, contended that Beijing was unable to replace Hong Kong as a global financial hub because environmental pollution makes it unable to attract and retain top talent in financial services.[34]

Environmental health issues also are contributing to potential food security problems in China. Among them, air pollution has a negative impact on grain yield.[35] In 2003, acid rain and sulfur dioxide pollution were estimated to have cost China 30 billion *yuan* ($3.62 billion) in agricultural output.[36] Surface ozone – a major component of smog – can penetrate the plant structure and impair its ability to develop and is thought to be the cause of reduced yields of summer wheat by 6–12 percent each year and soybeans by 21–25 percent.[37] A study aimed at quantifying the environmental and human health impacts of ozone and PM2.5 found that exposure to ozone alone destroyed 20 million tons of rice, wheat, maize, and soybean in 2010. The crop losses, if combined with economic cost of air pollution associated with premature deaths (1.1 million), shaved $38 billion annually off the Chinese economy.[38]

Research by Chinese scientists has also demonstrated that air pollutants can severely impede photosynthesis, the process by which plants use sunlight in the air to synthesize food from carbon dioxide and water.

A professor at China Agricultural University warned that if smog persists at high levels, China's agriculture sector would suffer conditions "somewhat similar to a nuclear winter."[39] Taking into account reduced agricultural output as well as pollution-caused absenteeism and increased medical spending, an OECD study predicts that air pollution may shave the size of China's economy as much as 2.5 percent by 2060.[40]

In order to reverse this perennial hemorrhage of national economic benefits, the government cleanup efforts have to catch up with the speed of pollution. Experts initially estimated that China should spend at least 2 percent of GDP each year on pollution control, roughly double the level of current spending and far above what the government was investing in the early decades of the post-Mao era.[41] Failure to spend that much, they argued, may cost the Chinese economy half as much again in "blighted crops, health costs and pollution-related expenses."[42]

THE DIVERSE EFFECTS OF POLLUTION

Pollution in China is ubiquitous, but it also has widely varying effects on people in different income groups. In examining the health risks associated with water pollution, two Chinese economists noted that those with low socioeconomic status were in a more disadvantageous position to reduce exposure and protect their health.[43] Research has also shown that people's ability to invest in goods to protect themselves varies across income groups. Purchases of air filters are common among all income groups, but the rich and middle class are much more likely than the poor to buy more filters in response to worsening pollution.[44] In heavily polluted urban centers, wealthy residents also have other ways to increase their access to clean food, water, and air. For example, they may fly to China's southern island province, Hainan, for respite from debilitating pollution. In December 2016, when Beijing issued a red alert on smog, the city led searches in the nation on the travel website Qunar.com for "avoiding smog," "wash your lungs," and other terms related to travelling, while searches for flight tickets from the city to coastal locations such as Hainan quadrupled.[45]

The cost of environmental health is not evenly distributed across regions, either. Absent effective intervention, provinces or municipalities with a high PM2.5 concentration (e.g., Tianjin, Shanghai, Beijing, Jiangsu, Hebei, and Shandong) incur more GDP and welfare losses than those with relatively low PM2.5 concentration (e.g., Xinjiang, Inner Mongolia, Hainan, and Gansu).[46] But less-developed provinces with significant labor-intensive and resource-intensive industries are also more likely to suffer economically from the health consequences of environmental degradation.[47] Variation in healthcare provision may also matter. Better access to healthcare in the developed eastern provinces helps mitigate the negative effects of intensive pollution; lower quality healthcare in the less developed western provinces may exacerbate pollution's negative health effects.

While China's growth is still continuing at a relatively high level, compared to most other nations, the economic effects of environmental health challenges are threatening its sustainable development. According to President Xi Jinping's political report at the 19th Party Congress, in October 2017, China will "basically realize socialist modernization" by 2035. The new objective is much more ambitious than the one set by Deng Xiaoping three decades before, which anticipated China would achieve modernization by the middle of the twenty-first century. Although Xi did not specify growth targets in the report, it is estimated that in order to fulfill the new goal, China's GDP per capita has to reach more than $20,000 (in present value), compared to $11,000 today, and sustain an annual growth rate of at least 5 percent between 2020 and 2035.[48] But China faces a so-called ecological trap: it must devote a significantly larger share of GDP to pollution control or contend with the even greater economic costs of further damage to its environment. Either way, the consequences may hinder economic growth and increase the danger of China falling into the middle-income trap, which occurs when a country remains stuck at a level of economic activity per person well short of advanced economies.[49]

Here's how Ha Jiming, an esteemed Chinese economist, described the challenge:

> Environmental protection is an additional structural problem or a fault line [in China's economy]. Air pollution, water resources pollution, even soil pollution will all constrain China's future economic growth .... Because

a large proportion of annual GDP will be spent on fighting pollution, there will be actually not much (economic) growth. If economic growth is premised upon resource destruction, that growth is not sustainable. China is now indeed in such a dilemma, with so many problems and challenges.[50]

ENVIRONMENTAL HEALTH AND SOCIOPOLITICAL STABILITY

Environmental health issues not only exact a significant economic toll but also have profound social and political implications. Ever since 1989, when the government brutally cracked down on the student democracy movement, social-political stability has been the dominant concern of ruling elites and indeed the main driver of public policy dynamics. It is no coincidence that since the 1990s, there has been a growing number of public protests, petitions, and civil disobedience over grievances ranging from corruption to environmental degradation. The number of annual protests ("mass incidents") increased steadily from approximately 8,700 in 1993 to 180,000 in 2010.[51] The government's obsession with social-political stability explains why since 2010 China's spending on domestic security, including policing and surveillance, has consistently outstripped military spending.[52]

As University of Chicago professor Dingxin Zhao pointed out, essentially there are three ways that state power is rationalized: legal–electoral, ideological, and via performance.[53] Since the Chinese regime is not based on popular elections or rule of law, and can no longer justify its rule by self-praised ideological superiority, the state has to deliver good performance to an increasingly demanding public. True, even democratic governments have to worry about their performance – but the main consequence in a well-functioning democracy is that they simply lose the next election. They still live to fight another day.[54]

Unlike in a democracy, people in China have no effective, institutionalized means to hold government officials accountable. They are instead stuck with the government even if they are not pleased with its performance. In this top-down and state-dominated political structure, bad performance by government officials often is considered the equivalent of bad performance by the regime itself. When conducting fieldwork in China, I was amazed that so many people were comfortable attributing a policy failure or a social-economic problem to the failure of the "system" (*tizhi*).

While the post-Deng Xiaoping leadership can rely on economic growth and nationalism to shore up its political legitimacy, the absence of a solid ideological consensus and rising public expectations make the social contract between the leaders and the led more tenuous than it was in the Mao era. A major economic setback or the inability of the government to respond to people's legitimate demands could evolve into a sociopolitical crisis. Performance-based legitimacy is therefore intrinsically fragile. President Xi was keenly aware of the vulnerability of this type of legitimacy when he remarked in late 2013 that "If we do not do a good job in food safety, and continue to mishandle the issue, then people will ask whether our party is fit to rule China."[55] Since 2013, with the growing public attention on air quality, pollution has become a political issue that tests the Chinese government's ruling capacity.[56]

SOCIAL RESPONSE TO POLLUTION: EXIT, VOICE, AND LOYALTY

To the extent that such a "defensive regime" has to base its survival on performance-based legitimacy, environmental health issues surely will threaten political rule through their impact on economic growth. The stakes are even higher given Beijing's well-placed fears over how the Chinese public has chosen to respond to environmental health issues and the potential for more explosive reactions in the future.

The framework developed by Albert Hirschman, in his description of citizens' responses to deteriorating organizations and situations, provides a useful tool for analyzing the situation in China. He made a distinction between three alternatives: Exit (quitting the organization or denying the organization their loyalty); Voice (agitating and exerting influence for change from within); and Loyalty (accepting and adapting to the deleterious change).[57]

Loyalty is the most common initial response. Feeling powerless, individuals and families may choose to live with the effects of environmental degradation, although they do make individualized efforts to minimize the harm, including purchasing air purifiers, drinking bottled water, and consuming imported rice. According to statistics gathered by the Chinese online shopping website Taobao, in 2013, when smog became a major health concern, Chinese consumers spent 870 million *yuan* ($141 million) on 4.5 million online transactions purchasing antismog products, increasing the number of customers who bought facial masks and air purifiers by 181 percent and 131 percent, respectively, from 2012.[58] A survey of 1,050 Beijing residents in April 2015 found that 81 percent of them had taken certain individual steps to reduce the health hazards of air pollution, with "use of air purifiers at home" and "increasing the number of green plants at home" being two of the most common measures.[59] Taking food safety into their own hands, farmers came to trust only the produce they grew on their own farmland, while consumers picked vegetables with holes chewed in them (as a sign that pesticides were not used), and avoided small grocery stores (which are considered more likely to sell unsafe and low-quality products) when shopping. Such adaptive responses by individuals to a toxic environment – what Andrew Szasz calls "inverted quarantine"[60] – do not pose any fundamental threat to regime legitimacy.

The question is: how long will Chinese consumers tolerate the status quo, or, to borrow from mathematics, where is the inflection point? State counselor Liu Yanhua warned about the sociopolitical consequences of air pollution:

> Smog … is caused by the economic development model, and failure to address it will result in social problems. Imagine how much and how long the society can tolerate such severe air pollution? If we copy the experience of the West, where it took more than 10 or even 20 years to clean up, a large number of people cannot accept [that long to fix the pollution]. The expectation that hundreds of millions of people's health will be severely harmed cannot be tolerated.[61]

Rather than take things as they are, some well-to-do Chinese choose to "exit" or emigrate. One study finds that almost half of wealthy Chinese want to emigrate.[62] The number of emigrants from China – mostly wealthy citizens – more than doubled, from 4.1 million in 1990 to 9.3 million in 2013.[63] Pollution increasingly is becoming a primary reason driving this trend. Research has shown that a 100-point increase in the air quality index (AQI) led to a next-day rise in Internet searches of the term "emigration" by approximately 2.3–4.8 percent, and that the searches would go up further when the AQI increased by more than 200 points.[64] Several years ago, I was visited by one of my high-school classmates, whose family had just emigrated to Canada. He made no secret of the fact that pollution was the primary reason for their departure from China. They are not alone. According to a survey conducted by Shenzhen-based *New Fortune Magazine*, more than 70 percent of wealthy respondents identified the quality of the environment and healthcare as important factors in the decision to emigrate.[65] A 2014 report published by the Center for China and Globalization (CCG), a nongovernmental think tank in China, showed that concerns about air and water pollution as well as food safety were driving elites and wealthy Chinese to leave the country.[66] In a more recent survey, 56 percent of Chinese millionaires with a net worth of more than $1.5 million were considering moving overseas or have already immigrated to another country. Education and environment were cited as the two main reasons driving them to emigrate.[67]

While emigration as a way to escape pollution is an option for only a small segment of the population, the lack of confidence among the country's economic and educated elite should concern CCP leaders, as their departure would result in a dramatic loss of earning power and a potential brain drain.[68] The exodus of these people – generally between 35 and 55 years old, represents a loss of both human and financial capital. In 2012, for example, more than 6,000 Chinese who left for the United States took an estimated $3 billion to $6 billion with them.[69] The CCG report estimated that China's wealthy – those with personal investable assets of more than 6 million *yuan* ($0.93 million) – had allocated a combined $433.5 billion assets overseas by 2012.[70] The total number of Chinese receiving the EB-5 visa granted to foreigners who invest $500,000 in a US-based development project reached 8,500, an all-time high, maxing out the program's yearly quota.[71]

In comparison to Loyalty or Exit, the Voice option has more profound implications for sociopolitical stability. In exercising the option to speak out for redress of environmental health–related grievances, the Chinese people have shown a clear lack of interest in judicial means to amplify their voices. During the period from 2006 to 2010, for example, there were only 1,010 environment-related administrative and criminal litigation cases, representing less than one percent of environmental disputes. During the same period, government authorities received more than 300,000 environment-related petition letters and 2,614 petitions asking that administrative cases be reconsidered or reheard.[72] Indeed, between 1995 and 2010, the number of environment-related petitions increased more than tenfold, from 58,678 to 701,073. The number of the letters fell in 2011, probably due to the growing use of phones and the Internet. If letters, phone, and online complaints are combined, the number of petitions and complaints has continued to increase. Between 2011 and 2015, petitions grew by more than two-thirds, from 1.05 million to 1.77 million (Figure 2.1). "Environmental petitioners" are now a recognized subset of the huge population of petitioners in China.

While letter writing or phone campaigns are indications of rising awareness of environmental and rights issues, they rely – just as in filing

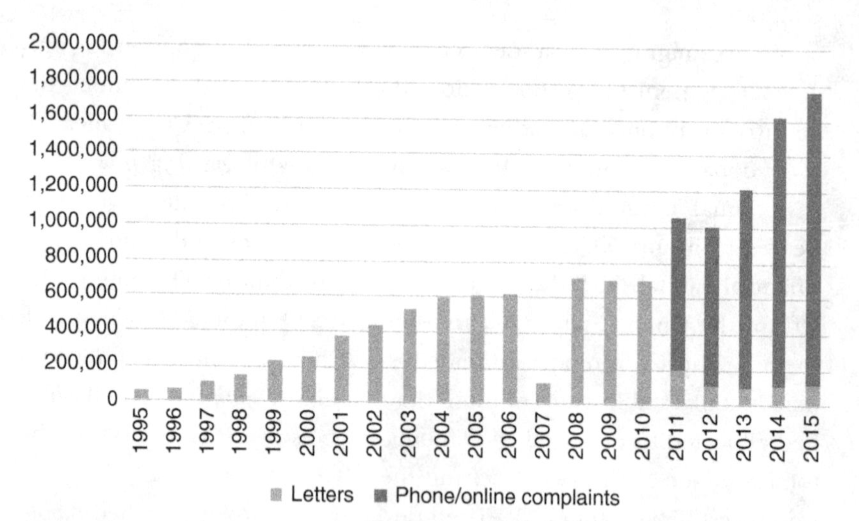

Figure 2.1 Number of environment-related petition letters and complaints, (1995–2015). *Source*: Tong Zhifeng, "Licheng yu tedian: shehui zhuanxing qi xia de huanjing kangzheng yanjiu" (Processes and Features: Studies on environmental protests in social transition periods), *Gansu lilun xukan* (Gansu Theory Research), no. 6, November 2008, pp. 85–90; MEP, *Quanguo huanjing tongji gongbao* (Statistical bulletin on national environment), 2010–2015.

a complaint with a court – primarily on government-sanctioned means intended to encourage reform within strict guidelines set by the state. The voices and concerns of the people are channeled into a political system that institutionalizes an asymmetrical power relationship between the state and society, in which the state enjoys near monopoly on formal power, with minimal checks and balances in policymaking. Industry and business, which share an interest in starting and sustaining projects for profit, are the natural allies of local government officials who have strong incentives to pursue economic and revenue growth in their jurisdictions.

In contrast to the powerful and stable local government–business alliance, residents affected by environmental degradation not only are deprived of the right to participate effectively in policymaking but also are unorganized and unable to protect their interests through the channels shaped by the state. When their patience runs out, they have no choice but to turn to informal, unconventional, even illegal means to mitigate perceived risks or hazards. From 2000 to 2004, there were a total of 16,523 mass incidents in Guangdong province alone, 62 percent of them triggered by the government's failure to respond effectively to people's complaints.[73] In view of the cozy relationship between government and business, their actions and campaigns against the polluting projects ultimately led them to press for government interventions and solutions. Such "mass incidents" take the form of spontaneous or planned gatherings, traffic obstruction, encirclement of government compounds, sit-in protests, mob attacks, and other collective actions. According to Yang Chaofei, an MEP chief engineer, the number of mass incidents protesting environmental conditions grew an average of 29 percent a year from 1996 to 2011.[74] In 2005, Zhou Shenxian, the Minister of Environmental Protection, acknowledged 51,000 pollution-related protests, or roughly 1,000 a week.[75] By 2012, pollution rivaled illegal land acquisition and labor disputes as the leading reason for mass incidents in China.[76]

The difficulties Chinese people face and their determination to confront polluting firms were vividly captured in *Warriors of Qiugang*, a Yale Environment 360 documentary co-produced by Ruby Yang and Thomas Lennon.[77] In 2004, a chemical plant was set up in Qiugang, a small village in China's eastern Anhui province, to produce pesticides and dyes.

Chemicals churned out by the firm soon turned the local river black, contaminated grain crops, sickened children, and led to unusually high outbreaks of cancer in the village. Local people who sought redress confronted the factory, filed lawsuits, and petitioned upper-level governments, only to find that government officials colluded with the factory owners and showed no interest in shutting down the factory. Disappointed, they staged a more violent protest, assaulting and beating the party boss of the township. To calm the villagers, the government finally agreed to relocate the factory by the end of 2008.

Pollution-related health concerns are an important trigger of environmental protests and activism in China. Deputy Minister of Environmental Protection Wu Xiaoqing remarked at a 2012 conference that of all the significant environmental incidents during 2006–2010, a quarter of them involved health problems. Wu additionally noted that of all environmental health incidents, 80 percent had occurred in the countryside, 89 percent were concentrated in the chemical industry and metal mining, smelting and recycling industries, and more than half evolved into mass protests.[78] From 2013 to 2017, 712 local anti-pollution demonstrations were officially recorded, although other estimates suggest as many as 30,000, or even 50,000 took place.[79]

Why, given so many other serious problems that contribute to social discontent – among them corruption and income inequality – are environmental health issues such an effective mobilizer for mass protests in China? Because they strike directly at people's own lives.[80] According to Liu Jianqiang, a former investigative journalist in China, environmental pollution crosses a red line by threatening health and life. In his words, "one can have a factory owner as a neighbor who is 10,000 times wealthier than he is, but he cannot tolerate his child getting sick because of the pollution produced by that factory."

In a similar vein, Daniel Gardner noted the importance of connecting environmental contaminants directly to public health:

> When, for example, "smog" was just "fog," and an irritant to the throat, nose, and eyes, the public found it inconvenient and unpleasant, but their tolerance was higher. However, once "fog" became PM2.5, and once it was shown that

> PM2.5 could lead to death and reduce life expectancy by more than five years, dirty air became a far more personal affront requiring urgent action.[81]

There is another reason environmental protests have become so frequent: because they do not necessarily challenge the authority of the existing regime publicly, the risk of reprisal and punishment by the government is relatively low. Furthermore, unlike illegal land acquisition or labor disputes, an environmental problem is often widespread: a large-scale chemical industrial firm, for example, may impact more than a million people's health. Once a sufficient number of people consider themselves harmed by pollution, there is more safety in numbers and more people are likely to overcome their fear and take to the streets.[82]

AN ANALYSIS OF ENVIRONMENTAL HEALTH–RELATED MASS INCIDENTS. Table 2.1 documented thirty-two environmental health–related mass incidents reported by domestic and/or international media between 2005 and 2017. They vary in terms of size, location, and patterns of protest participation. About half of them occurred in rural areas, where villagers protested against pollution from ongoing or proposed local industrial projects. Most of the remaining cases feature urban residents taking to the streets protesting chemical or waste disposal plants slated for construction in their cities. While thirteen incidents (1–3, 5, 7–10, 12, 13, 15, 22, and 25) sought redress after the damage was done ("*ex post* responses"), the remaining incidents targeted proposed projects and demanded the government adopt preventative measures ("*ex ante* actions").[83] Such *ex ante* actions began with the Xiamen PX incident in 2007, when nearly 10,000 residents of the coastal Fujian city marched in the streets to protest the planned construction of a local petrochemical plant. Proactive environmental protection has risen since 2012. Compared with *ex post* responses, *ex ante* actions are relatively easy because powerful interest groups, such as the government, employers, employees, and their family members, have not yet formed around existing polluting projects, while the opposition to projects not yet put into operation is more likely to coalesce into a common front.[84]

TABLE 2.1 *Major environmental health–related mass incidents (2005–2017)*

Year	Mass Incidents	Health Concerns
2005	1. Dongyang, Zhejiang: reports that two elderly women were killed during a protest against local factory pollution sparked a bloody riot by thousands of villagers.	food safety; toxic gas leakage; pollution-caused illness
	2. Xinchang, Zhejiang: 15,000 villagers massed at a pharmaceutical plant and turned violent.	local pollution and its health effects
2007	3. Cenxi, Guangxi: villagers protested against a local paper mill for causing pollution.	pollution-associated food safety
	4. Xiamen, Fujian: nearly 10,000 urban protestors rallied over a proposed PX plant.	danger of cancer and birth defects caused by exposure to paraxylene (PX)
2008	5. Huaping, Yunnan: villagers demanded compensation for air and water pollution caused by a local cement factory.	food safety associated with dust produced in manufacturing
	6. Shanghai: Hundreds of residents took to "strolling," a wry form of protest against a planned extension of the city's maglev.	illness caused by radiation emissions
2009	7. Jiangyin, Jiangsu: villagers protested against a container shipment company for air pollution.	health issues caused by iron ore powder
	8. Liuyang, Hunan: villagers protested against a chemical industrial firm for cadmium pollution.	heavy metal pollution
	9. Fengxiang, Shaanxi: villagers stormed a factory.	lead poisoning of children
	10. Quanzhou, Fujian: villagers kidnapped local cadres for not taking actions against water pollution at a local factory.	food safety and health problems associated with water pollution
	11. Guangzhou, Guangdong: local residents protested against a proposed refuse-burning plant.	Dioxin released from burning refuse
2011	12. Haining, Zhejiang: villagers protested against a solar industrial firm for air and water pollution.	cancer villages
	13. Dalian, Liaoning: tens of thousands of residents protested against a proposed PX plant.	typhoon to breach chemical storage tanks and flood the city with toxic material
	14. Shantou, Guangdong: villagers protested against a proposed power plant.	cancer villages
2012	15. Shifang, Sichuan: thousands of people protested against a copper plant that they feared posed environmental and public health risks.	heavy metal pollution
	16. Qidong, Jiangsu: 20,000 demonstrators, fearing environmental degradation and health issues caused by a proposed wastewater pipeline project, occupied a government building.	carcinogenic effects of wastewater
2013	17. Anning, Yunnan: peaceful demonstration against a polluting PX chemical plant slated for construction in the city.	carcinogenic effects of PX
	18. Heshan, Guangdong: local residents took to the streets demanding a proposed $6 billion uranium processing plant be scrapped.	illness caused by radiation

TABLE 2.1 *(continued)*

Year	Mass Incidents	Health Concerns
2014	19. Maoming, Guangdong: local residents protested against a proposed PX plant and clashed with police.	carcinogenic effects of PX
	20. Luoyang, Guangdong: several thousand local residents protested against a planned waste incinerator.	toxic and carcinogenic effects of refuse burning
2015	21. Youxi, Fujian: villagers protested against the construction of a hazard disposal and recycling project.	medical waste; toxic waste
	22. Zhenping, Henan Province: thousands of local residents attacked a pharmaceutical factory that had been emitting polluted water and gas for 10 years.	toxic water and air released from the factory
	23. Luoding, Guangdong: 1,000 villagers rallied to protest against a proposed waste incinerator project.	health of future generations
	24. Jinshan, Shanghai: thousands of local residents took to the streets to protest against a proposed PX plant.	toxic air pollutants
	25. Shangrao, Jiangxi: thousands of villagers marched to a neighboring industrial park to protest.	emission of toxic water and air
2016	26. Xiantao, Hubei: tens of thousands of villagers protested against the proposed construction of a waste incineration facility near their community.	dioxin released from burning refuse
	27. Zhaoqing, Guangdong: villagers demanded authorities scrap plans to build an incinerator plant, fearing the plant would contaminate a nearby drinking water source.	dioxin released from burning refuse
	28. Xi'an, Shaanxi: 10,000 people gathered outside of government offices and clashed with police over a waste incinerator plant planned for their neighborhood.	cancer villages
	29. Lianyungang, Jiangsu: thousands of residents took to the city streets to oppose the building of a nuclear fuel recycling project.	illness caused by radiation
	30. Qianjiang, Hubei: Tens of thousands of local residents took to the streets in protest over a planned pesticide factory.	water pollution by pesticide factory
2017	31. Daqing, Heilongjiang: hundreds of residents protested against the proposed building of an aluminum processing plant.	hazardous substances released in aluminum processing
	32. Qingyuan, Guangdong: thousands of local residents took to the streets to protest against a proposed waste incinerator project.	water pollution and food safety

Source: Author's database.[85]

In contrast to earlier eras, when protests focused more on material losses, since 2012 people have been increasingly demanding the right to a good environment and decent health.[86] With rare exceptions (6, 17), most of the protests evolved into violence that led to government intervention and ended up with officials catering to the demands of the protestors. The number of people participating in such mass incidents varied, from several hundreds to tens of hundreds. Six of the incidents (2, 13, 16, 26, 28, and 30) involved more than 10,000 protestors.

These cases also have a lot in common in terms of the causes, context, conflict pattern, and outcomes. They all were driven by deep-seated fear of the health effects of pollution. The protestors explicitly linked ongoing or planned industrial projects to cancer villages (12, 14, 28), public health risks posed by heavy metals (8, 9, 15), and illness caused by radiation (6, 18, 29). In the case of Xinchang (2), villagers' initial demands included free physical and medical care for residents who live near the polluting plant.[87] They essentially were "not in my backyard" (Nimby) protests by local residents aimed at protecting their communities from the negative impacts of industrial facilities.[88]

In doing so, they share a foundation of accumulated anger at the failure of the local government to engage the public in the policy process. The polluting projects usually were launched without taking into account local residents' wants and interests. Public hearings on the projects, if held, often were a mere formality. The decision-making process usually was murky and exclusive so that local people were not only poorly informed about the project but also deprived of adequate time to voice their concerns. When the perceived health effects became common knowledge, local government officials often failed to respond to public demands raised through legitimate and institutionalized channels. In at least eight cases (1–3, 7–9, 12, and 22) the polluting plant in question had been operating under dispute for more than three years.[89] In the case of Zhenping, Henan Province (22), residents had protested for ten years against the pollution-emitting drug factory without success. Fed up with government inaction, the protesters often directly confronted the polluting firm in question, only to find that the local officials sided with the polluters. When intermediary institutions (e.g., NGOs, the media, and human rights lawyers) are not available for help, citizen protests often become "isolated activism"[90] and can escalate quickly into violence against the local government, frequently bringing on the intervention of higher levels of government, even central leaders. In that sense, the growth of mass protests at environmental conditions should be considered a genuine crescendo of voices rising in reaction to the failure of local governance.

As the party-state places a great premium on social stability, local government leaders typically face high levels of pressure to back down and end crises quickly. As seen in the water pollution riot in Qidong, Jiangsu Province (16), protesters stormed government buildings, smashing computers, overturning desks, exposing expensive liquor and condoms in government offices, throwing government documents out of windows to loud cheers from the crowd. The city's party boss was forced to strip down to his underwear while the mayor was forced to wear an opposition T-shirt.[91] Once the protest turned violent, not only did the local leaders temporarily tolerate being humiliated by protestors, but they also immediately announced that they would suspend the wastewater pipeline project.[92] Since the 2007 Xiamen PX incident, almost all the proposed PX projects have died out, and a growing number of refuse incineration projects have been shut down.

At first glance, the rise in people power seems to be a positive development in today's China. But such protests "succeeded" not so much because of the institutionalized elevation of people's voices in the public policy process but because of the government's ad hoc response aimed at tamping down unrest and maintaining sociopolitical stability.[93] The protest outcome is consistent with earlier findings that none of the protests had resulted in serious policy change.[94] Chinese environmentalists are concerned that the protestors' success in one case simply alerts the authorities to ramp up prevention and control measures against future protests, generating more difficulties for citizen movements in other localities.[95] State control in this area is facilitated by a rapidly expanding, "near-total" national surveillance network that features more than 176 million public and private surveillance cameras and sophisticated facial recognition technology.[96]

Yet, in a strategic environment, the government obsession with stability also encourages Nimby campaigns to grow because the government is more likely to respond to the protestors' demands the more they escalate into violence. This creates a vicious cycle: under the existing governance and policy structure, Nimby campaigns tend to grow large quickly and a "victory" by protestors then emboldens others with similar grievances to follow suit. In consequence, environmental mass incidents become "contagious" – there were five mass incidents involving PX projects (4, 13, 17, 19, and 24) and eight involving a waste disposal project (11, 20, 21, 23, 26,

27, 28, and 32). The pro-government *Global Times* lamented this vicious cycle, contending that the victory of the Xiamen residents in the 2007 PX incident "set a bad precedent for the public's unreasonable rejection of heavy chemical projects."[97]

So even though such Nimby campaigns are isolated and localized events, put together they can be viewed as part of a nationwide, coherent movement whose dynamics increasingly test the capacity of the Chinese state to govern. As a professor of China's Central Party School reminds us, environmental protests over the past decade have already been the most powerful social movement in China.[98] Keenly aware of their threats to sociopolitical stability, Xi Jinping emphasized, "the ecological environment is a vital political issue that pertains to the party's missions and goals, and also a vital social issue that pertains to people's well-being."[99] A government document highlights this aspect of the environmental health problems in China:

> Environmental health is both a complicated scientific problem and a sensitive social problem about which people are highly concerned. It matters to societal harmony and stability, the state's lasting political stability, and national survival and multiplication.[100]

ENVIRONMENTAL HEALTH AND FOREIGN POLICY

THE INTERNATIONAL ORIGIN AND SPILLOVER EFFECTS OF POLLUTION. China's environmental health problems are primarily driven by the interaction between its impulse for economic development and its domestic political structure. But the problems are not entirely endogenous. International trade and investment are also significant contributors to environmental degradation in China.

Consider rare-earth elements (REE), which are naturally occurring materials commonly used in high-tech devices and military equipment. REE mining requires large amounts of carcinogenic toxins and often leads to exposure to cadmium, lead, and radioactive materials. Because of concerns about the cost of production and health risk, the United States, Europe, and Japan, beginning in the 1990s, largely ceded the production of REE to China, enhancing its economic sway. But now China bears most of the environmental and health effects of its control.

Processing one ton of rare earths produces an estimated 2,000 tons of toxic waste, and each year the sector is churning out more than 22 million tons of toxic waste.[101] The toxic aftermath of rare earth mining with which China has to wrestle is indicative of a much bigger problem in China's export sector. In 2006, China's entire export sector was associated with 36 percent of the nation's sulfur dioxide (SO_2) emissions, 27 percent of its nitrogen dioxide (NO_2) emissions, 22 percent of carbon monoxide (CO) emissions, and 17 percent of the black carbon emitted in the country.[102] On top of that, export-related emissions in China contributed to 15 percent of annual average PM2.5 concentration in 2007, causing 157,000 deaths or 12 percent of the total global mortality attributed to PM2.5.[103] In 2010, China's export sector accounted for 1.8 million tons of PM2.5 produced in the country – 38 percent higher than the annual emissions in all EU countries – and 60 percent of these emissions in exports can be attributed to consumption in OECD countries.[104]

As more advanced industrialized countries sought to reduce their own air pollution, one of the ways they hit their goals was by outsourcing heavy manufacturing to China. But now, countries as far away as the United States are discovering that they can't necessarily escape the consequences of those actions. China is the main contributor to the planet's largest mass of PM2.5 air pollution, which can stretch from Mongolia to the Yellow Sea and often as far as South Korea and Japan. Based on high-resolution satellite data, NASA scientists found that local NO_2-emission-control effectiveness in Japan and South Korea may have been diminished by increasing transboundary transport of the NO_2 pollution from China.[105] The findings are consistent with the results of a recent joint research project by American and South Korean scientists who analyzed the contribution rate of PM2.5 in Seoul's Olympic Park. They found that China contributed to at least one-third of the site's PM2.5 pollution, most of it emanating from Shandong Province.[106] The study may well have downplayed the contribution from China because it was conducted in the summer, when China's PM2.5 level was relatively low.

Similar research has identified air pollution from China and several other Asian countries as major contributors to smog in the Western United States, accounting for as much as 65 percent of the region's ozone increase, compared to less than 10 percent from wildfire emissions and 15 percent from methane.[107] On days when strong winds are blowing across the Pacific, pollutants of Chinese origin contribute 12 to 24 percent of sulfate concentrations, 4 to 6 percent of carbon monoxide, and up to 11 percent of black carbon pollution in the Western United States.[108] NASA physicists suspect that large swaths of air pollutants in China and other Asian countries, when blowing east, may mix with storms above the Pacific Ocean and contribute to colder and snowier winters in the United States.[109]

IMPLICATIONS FOR GLOBAL GOVERNANCE. The health effects of China's degrading environment also have profound implications for global health governance. Many major disease outbreaks, including the 1957 Asian flu, the 1968 Hong Kong flu, and the 2003 SARS epidemic, originated in China and affected the patterns of disease outbreak elsewhere. Like these infectious disease outbreaks, transboundary pollution is becoming a killer worldwide. As early as 2000, airborne pollutants from China were responsible for approximately 30,000 additional deaths outside the country.[110] The impact becomes even larger over time. *Nature* magazine published a study estimating that 30,900 premature deaths in Japan and South Korea in 2007 were related to PM2.5 pollution in China.[111] As a leading scholar on China's environment noted: "If China does not act on its environmental problem, then Japan and South Korea will bear the consequences."[112] In addition, globalization allows contaminated food from China to enter the food supply chains of other countries, as evidenced by the discovery of Chinese-made dumplings tainted with the insecticide methamidophos in Japan in 2008 and Chinese tinned peaches high in lead in Australia in 2014.[113] Indeed, food safety concerns, like heavy metal concentration in Chinese poultry, have been a main reason behind the US ban on imports of chicken products from China.[114]

The United States did not face a major environmental health crisis in its pre–World War II emergence as a superpower. China does not enjoy such a luxury. Its enormous environmental health problems represent not only a challenge to its global ambitions but also an opportunity to demonstrate more enlightened leadership. On the one hand, China's environmental health problems, in conjunction with other mounting domestic challenges, will constrain Chinese leaders' ability to mobilize the resources and internal support necessary for Beijing to challenge the United States on the world stage. On the other, to the extent China succeeds in mitigating its domestic environmental damage, it could help reduce disease and ill-health around the world and win credit abroad for a more enlightened environmental policy.

This situation represents an opening for a strategic and environmentally sensitive trade policy on the part of both China and the United States. Since the agriculture and steel industries are major contributors to environmental degradation in China, catering to US demands for more balanced trade by increasing imports of farm products from the United States (e.g., poultry and pork) and reducing the exports of steel products to the United States would improve the bilateral relations while having a significant impact on reducing emissions in China.[115] Furthermore, by cutting the excess and backward capacity in heavy industries, it would upgrade China's industrial structure. Although doing so may slightly increase the level of air pollution in the United States, the overall impact on the global environment would be positive because the United States has more effective environmental regulation and a greater capacity to minimize any damage.[116]

A domestic attack on Chinese pollution offers another advantage on the global front. Since toxic pollutants and greenhouse gas emissions have the same primary sources (burning fossil fuels), tackling environmental health issues may generate sufficient public support in China to closely work with other countries in addressing climate change and other pressing global environmental challenges. Scholars have observed that domestic and international concerns about air pollution and government commitment to pollution control have increasingly become the new driver for low-carbon development and facilitated the shift toward more effective climate governance in China.[117] According to Chang Jiwen, a senior researcher at a think tank affiliated with the State Council (China's cabinet), in drafting China's response to climate change prior to the 2015 United Nations Climate Change Conference (COP21) Chinese policy makers strategically were using the PM2.5 crisis to press for significant policy change around climate issues.[118]

President Xi himself said that addressing climate change was among the "intrinsic requirements" of achieving sustainable development in China.[119] Driven by this enlightened self-interest, China abandoned its previous rhetoric – that it had the right to use dirty technology just as wealthy nations had long done – and behaved instead as a leader and convener for a serious climate change agreement at COP21. In his address to the United Nations, Xi Jinping effectively combined the concerns of environmental health with those of climate change by talking about the importance of "natural resources such as air, water, soil and blue sky," arguing: "We must maintain harmony between man and nature and pursue sustainable development."[120]

At the same time, the presence of China's environmental health issues as a negative externality may prompt affected countries and other stakeholders to join the global battle against pollution. As two American environmental health scholars noted, when it comes to air pollution and health, we are actually in a worldwide fight to address a universally held concern, because "Earth's atmosphere knows no national boundaries, so the threat of airborne toxins is logically and constantly global."[121] In September 2014, the US Department of State hosted an inter-agency roundtable on PM2.5 air pollution in East Asia tasked with identifying ways to collaborate with China. The roundtable attempted to outline priorities for China in its engagement with its East Asian neighbors on particulate issues. Three years later, environmental ministers from China, South Korea, and Japan met and issued a joint statement pledging to step up efforts to tackle air pollution, climate change, and the loss of biodiversity.[122]

In the meantime, President Trump's "America First" approach to foreign policy and the US withdrawal from the Paris climate agreement have emerged as clear obstacles to US–China cooperation over environmental health issues. But they also give China the opportunity to position itself as a global standard-bearer and pacesetter on the environment, climate change, and sustainable development. In contrast to the Trump administration's opposition to controlling climate change and environmental hazards, Chinese leaders have pledged that China will not shirk its "international responsibility" to slash its emissions and will uphold the Paris Agreement.[123] China's stand over the Paris Agreement is consistent with the idea of a "community of shared future for mankind," which President Xi has been promoting since 2012. Likely derived from the Chinese concept of *tianxia* or "All-Under-Heaven," the new concept describes a world defined by mutual cooperation based on shared interests and responsibilities.[124] Such developments hold the promise of China emerging as a responsible leader in providing global public goods and promoting international cooperation, thereby mitigating the so-called "Kindleberger Trap."[125]

That said, China's role in global environmental leadership ultimately will be determined by its ability to fulfill its promises at the local level. According to a recent government report, almost 70 percent of the 14,000 companies examined by environmental inspectors failed to meet environmental standards for controlling pollution, calling into question whether China can indeed fill the global leadership gap on issues of environment and climate change.[126] Moreover, some of Beijing's new foreign policy initiatives that may reduce pollution and improve environmental health in China may exacerbate environmental health problems in other countries. One of the objectives of China's Belt and Road Initiative, for example, is to export surplus Chinese steel, cement, and other materials to smaller markets along the economic belt, which in itself is no different from the United States outsourcing large portions of its dirty manufacturing (and pollution) to China in decades past.

It is not hard to imagine who will ultimately bear the environmental cost of this initiative. "In 20 years, all that pollution will be in Africa," a provincial governor in China said.[127] The carbon dioxide gains from shifting to more advanced renewable energy technologies could be negated by the carbon dioxide produced in making concrete when China builds and constructs roads and airports in Belt and Road countries.[128] The initiative may also create new environmental health hazards in former Soviet republics and in Russia, where there is poor environmental governance.[129]

Even China's positive attacks on pollution pose complications as it seeks to enhance its global position.[130] Its new green policies subsidize auto manufacturers and tighten regulation over traditional internal combustion engines. Its heavy investments in the electric-vehicle market have already made China the global leader in electric-vehicle sales since 2015 and could undermine efforts in the United States and Europe to develop healthy electric vehicle industries of their own. China's huge investment in renewable energy as an alternative to coal has made China the world's largest solar panel maker, too. The sheer scale of its investment and fierce market competition have resulted in plunging solar panel prices, allowing a faster shift away from carbon worldwide. But China's solar offensive enables it to undercut foreign producers, set global standards and may allow it to corner the market in what is bound to be a major industry in coming decades.[131]

China's efforts to clamp down on polluting firms and cut excess industrial capacity have also had an enormous impact on international commodity prices. Closing factories and mines drove up the prices of coal, steel, and aluminum in 2017. Prices of tungsten, considered "one of the most critical strategic materials for the Western world's economy and defenses," went up by 50 percent in summer 2017.[132] Similarly, driven by concern for the environment and public health, China's ban on imports of "foreign garbage" (e.g., shiploads of other countries' plastic and paper trash) has caused a major disruption in the international recycling market, increasing the chances of recyclables in the United States ending up in landfills while at the same time driving up paper prices.[133] Environmental activists warn that the ban might lead to huge amount of waste being exported to some South Asian and Southeast Asian countries where waste industries are less well regulated.[134]

ENVIRONMENTAL NATIONALISM. China's leaders may recognize the importance of making strides on the environmental front, but they also have to be wary of an important element in the Chinese political landscape that is skeptical of shifting policy in that direction. In particular, negative attention on China's environmental health issues by foreign governments and the Western press has added ammunition to the arsenals of Chinese nationalists. A commentary published in *Global Times* criticized Western media as being hypocritical:

> In some Chinese perspectives, those who care most about China's environmental issues are not us Chinese but "civilized and progressive" Westerners. For example, the U.S. embassy, which measures PM2.5 levels in China, once was touted as an important promoter of China's environment protection ... But the facts have proved again and again that when touching upon their own national interests, Western countries do not really care about Chinese people's interests but will choose to defend their interests without hesitation ... They run around on sinister errands talking about those topics only because they are not required to sacrifice themselves.[135]

Many Chinese nationalists view international pressures for emission control as just another Western plot to contain China's rise. When doing fieldwork in Taiyuan, the capital of Shanxi province, I was told by a local government official that loud foreign demands on China – rather than India – to improve its environment was a plot, given his view that China did a much better job than India in protecting its environment.[136]

Zhou Xiaoping, an essayist and popular blogger who once was praised by President Xi for spreading "positive energy," accused Western countries of "sensationalizing smog topics to coerce public opinion to pressure Chinese government into signing the Framework Convention on Climate Change."[137] He further accused the US embassy of "spreading rumors" and using smog to vilify Chinese high-end major manufacturing firms so that China ultimately had no choice but to accept the low-end, low-profit, and highly polluting small firms outsourced from the United States.[138] This conspiracy theory, contradictory as it may seem, was embraced by some of China's leading intellectuals. In a video clip circulated widely on China's social media in January 2017, a top Chinese climate scientist and Vice President of Chinese Academy of Sciences, Ding Zhongli, was interviewed by Chai Jing, then a reporter with China Central Television (CCTV). Professor Ding defended China's unyielding positions at the 2009 Climate Summit in Copenhagen, arguing that China should be given the same "emission space" as Western countries. When Chai asked whether China should work with other countries in cutting as much emissions as possible, Ding responded: "Are Chinese humans? Why can a foreigner produce carbon emissions four times of a Chinese?"[139] While a very small number of intellectuals disagreed with Ding, support for his remarks flooded China's social media, with some calling him the "backbone" of the country.[140] In March 2018, Ding became vice chairman of the Standing Committee of National People's Congress (NPC), the supreme organ of state power in China.

The nationalists' objections to Western criticisms of China's environmental problems are intertwined with their hostility toward democracy and freedom. In May 2016, a Chinese student named Yang Shuping praised "sweet and fresh air" and freedom of speech in the United States in her commencement speech at the University of Maryland. Her speech circulated quickly on China's social media outlets. Rather than focus on the main message the speaker wanted to convey, most Chinese commenters magnified and overanalyzed Yang's statements and lambasted her for humiliating China. One commenter wrote: "The air in our country is bad, [but] this is not the problem. She is flattering Americans by saying our country is flawed. We are Chinese, between one another we can discuss what is wrong with our country, but we still love our homeland."[141]

Social media soon lit up with all kinds of conspiracy theories about Yang, the University of Maryland, and the university's president. A Chinese web portal featured a long story "revealing" the close ties between the university and US intelligence agencies and how the university has allegedly long engaged in "anti-China intelligence activities."[142] Appalled and disgusted, one leading China scholar said: "To put [Yang's] words to such a level of significance (humiliating China) only adds to stereotyping the Chinese and China that we all have fought to overcome."[143]

This rising environmental nationalism signals a change in official discourse around national identity regarding the West and has accentuated Chinese elites' growing fear of internal instability.[144] While Chinese leaders may not buy into these conspiracy theories, they might find foreign policy making in China increasingly constrained by irrational and aggressive environmental nationalism. Mitigating this foreign policy dilemma requires the government to map out an effective domestic public policy response to its environmental health-related challenges.

The question is: to what extent is the government ready and able to make such a change? This will be the subject of Part II of this book.

PART II

THE GOVERNMENT RESPONSE

CHAPTER 3

Evolving Environmental Health Policy

AFTER DECADES OF ASTONISHING ECONOMIC GROWTH THAT has turned China into the world's industrial powerhouse, is Beijing finally ready to take on the environmental degradation that has been its most damaging consequence? Chinese leaders have promised that they are.

But the experience of environmental activists like Chai Jing suggests that China may have lost the war on pollution even before it really begins. On February 28, 2015, Chai, a former China Central Television (CCTV) journalist, released her self-financed TED Talk-style documentary, *Under the Dome*, which investigated China's air pollution and connected the environment to health in an unprecedented way.[1] The film began with a story implying pollution was the cause of the tumor her unborn daughter developed in utero. Presented in a style similar to Al Gore's *An Inconvenient Truth*, Chai openly criticized the role played by state-owned energy, steel, and coal producers in contributing to the air pollution crisis and revealed how powerless the Ministry of Environmental Protection (MEP) was in fighting these industrial conglomerates. The film turned out to be a phenomenal success. Within three days of its release, it was viewed over 150 million times on Tencent, a leading provider of Internet-related services and products in China. The newly appointed Minister of Environmental Protection, Chen Jining (who was an academic before his promotion), personally thanked Chai for her work.

Chen even likened *Under the Dome* to Rachel Carson's path-breaking 1962 book *Silent Spring*, which documented the adverse effects of the indiscriminate use of pesticides, indicating that the MEP was hoping to use the film to galvanize support for a robust environmental movement in China. Public response to the film suggested he was not daydreaming. Many online viewers expressed their support for Chai. According to a survey, more than 80 percent of the viewers were deeply concerned about air pollution in China, about 70 percent said the film changed their views on smog, and over 75 percent expressed a willingness to change their transportation and energy use habits. Moreover, a strong majority said that not only industrial emitters but government officials should also be held accountable for the smog problem.[2]

Perhaps that is why party leaders soon had second thoughts about their support for *Under the Dome.* Within a week of its release, after another 150 million views were recorded, the video was blocked by government censors. The order to take the documentary offline allegedly came directly from Xi Jinping.[3] The scene of Chai receiving an award from Alxa SEE Ecological Association, an environmental protection organization initiated by some 100 well-known entrepreneurs, was also deleted from a TV program. Chai's last Weibo post came on January 1, 2016. Later that year, I dialed her cell phone number (provided by a journalist in China), trying to ask for an interview. No luck. She seemed to have completely faded away from the public eye. When facing the Beijing winter, the silent spring proved a flash in the pan.

Indeed, the contrast is stark when comparing the fate of *Under the Dome* with that of *Silent Spring.* Both strove to raise awareness of pollution's threat to human health. *Silent Spring* helped trigger the environmental movement in the United States, leading within less than a decade to the passage of the Clean Air Act and the establishment of the Environmental Protection Agency. In China, the government – apparently seeking to balance air cleanup with social-political stability – responded by tightening regulations on pollutants *and* on the public discussion of pollution control. After releasing their work, both Carson and Chai were vilified personally by special interest groups and their allies in and out of government. Carson, however, had the freedom to fight courageously for her cause, while Chai has been silenced, not given the chance to defend the truth of her film.

Given the magnitude and seriousness of the environmental health crisis, one has reasons to question the government willingness and capacity to confront the challenges in an effective manner. After all, the demand for policy change in this arena is not new – current calls for change might trigger a sense of déjà vu for anybody who is reminded of Premier Zhou Enlai's reasoning forty-eight years ago:

> The [environmental] problem I was worried about is indeed occurring, and is actually more serious ... We can no longer adopt a laissez-faire approach to China's environmental problem. It is time to put the problem on the government agenda![4]

The evolution of government policy toward environmental health is a long and winding tale. It highlights the roles of a host of players – political leaders, bureaucrats, the media, social forces, and international actors – in agenda setting and policymaking over environmental health. It also reveals the inherent dilemmas, constraints, and limitations China faces in pursuing required change.

RAISING AWARENESS (1960S–1990S)

To be fair, the Chinese Communist Party (CCP) has a tradition of tackling broadly defined environmental health issues. During the revolutionary era (1927–1949), Mao encouraged the mobilization of communist troops and local residents to keep their surrounding environment clean in order to minimize the risk of epidemics.[5] The interest in protecting environmental health was sustained in the early 1950s by the "patriotic hygiene campaigns" that emphasized improving environmental sanitation. By invoking the alleged threat of US germ warfare during the Korean War, the campaigns sent a strong message that environmental health was a national security concern. The environmental health work gained additional impetus when Mao became personally involved in eradicating schistosomiasis (a waterborne disease in southern China) and Ke-shan disease (a health problem caused by selenium-deficient soil). Meanwhile, the government issued factory safety and health regulations to protect occupational health. During the Great Leap Forward (1958–1961), Mao also launched a campaign against the "four pests" (mosquitoes, flies, rats, and sparrows). The party-state's venture into the domain of environmental health reduced the incidence rates of communicable diseases, which in turn legitimized the state's domination over society.[6]

These environmental health-related campaigns paid little heed, however, to the health issues associated with pollution. Within a few years of taking power in 1949, Mao was thrilled about the prospect of Tiananmen – the national symbol of China – being surrounded by factory chimneys.[7] Influenced by the Soviet development model, China's economic system then focused on state ownership, heavy industries, and central planning. This system lacked sufficient mechanisms to prevent the spread of public hazards and pollution. The Great Leap Forward and the Cultural Revolution (1966–1976), with their emphasis on human victory over nature, only wrought additional devastation.[8]

Between 1952 and 1982, for example, wastewater discharged from Jilin Chemical Co, one of China's largest chemical industrial enterprises, polluted the Songhua River in Northeast China. In consequence, cases of chronic mercury poisoning have been observed in the area since the mid-1960s. While the causes, symptoms, and timing associated with the outbreak were basically the same as those of the infamous Minamata disease outbreak in 1956 in Japan, the issue largely went unnoticed in China. Because of government cover-up as well as the widespread propaganda that socialist countries were free from pollution, even most policymakers in China thought China's environmental pollution, if any, was negligible and fundamentally different from public hazards in the West.[9] Prior to 1970, environmental protection remained an alien concept in China; "public hazard" (*gonghai*) – a term imported from Japan – was used in China to describe environmental pollution in Western countries.[10]

Premier Zhou Enlai was probably the only government leader who was keenly alarmed by the environmental mayhem in China. After two major water pollution incidents at Dalian and Beijing in the early 1970s, Zhou realized that public hazards were not unique to capitalist countries.[11] Between 1970 and 1974, Zhou made more than thirty-one speeches on environmental protection, covering issues from water pollution in Shanghai to air pollution in Beijing. One of them showed that he was very concerned about Beijing becoming smothered in smog like London:

> Beijing's air pollution is already very serious, and we should undertake immediate measures to address it! In the past people often mentioned "Foggy London." We may become "Foggy Beijing" if we do not do a good job [tackling pollution].

"SOFTENING UP" POLICY ACTORS TO NEW IDEAS. During the Cultural Revolution, however, a large number of bureaucratic leaders were purged and military representatives assumed control over government institutions. Consequently, even Zhou had problems finding an audience at the central government meetings when he wished to talk about environmental issues. To raise environmental awareness in China, Zhou in 1972 took an unusual step by sending a major delegation to the UN Conference on the Human Environment (UNCHE) in Stockholm, Sweden.[12] Given pollution's impact on human health, the initial plan was to send a small delegation led by the military representative of the Ministry of Health (then the de facto head of the Ministry).[13] Convinced that pollution was pertinent to many aspects of national economy, Zhou intervened, raising the number sent to Stockholm to more than forty delegates from the departments of agriculture, foreign affairs, health, light industry, metallurgy, national planning, nuclear industry, and petrochemical industry.[14]

The conference provided a great opportunity for the previously insular delegates to reexamine China's environmental problems with the aid of an outside perspective. Protestors attending the Conference shocked Chinese delegation head Qu Geping with images of disabled victims of pollution, sparking in him a new awareness of the severity of China's environmental problems. Back in Beijing, Qu summarized in his report to Premier Zhou that "the degree of pollution in Chinese cities and rivers is no less than in Western countries, but the degree of natural ecological damage in China is far more serious than in these countries."[15] Based on the report, Zhou in August 1973 organized China's first National Conference on Environmental Protection.[16] Zhou shared the meeting briefings with central ministers and provincial party secretaries, which helped soften up the "policy primeval soup" of ideas and urgency for solving environmental health problems.[17] After the meeting, the State

Council set up the Environmental Protection Leading Small Group. Environmental protection agencies, including departments conducting environmental monitoring and research, were set up from the center in Beijing down to the local level. A new consciousness among Chinese policymakers legitimized research on environmental health issues. After 1972, China began to monitor the concentration of the so-called floating dust – particles less than 100 micrometers in diameter, which ultimately led to research on PM10 and PM2.5.[18] The State Council Environmental Protection Leading Small Group also drew up a ten-year plan to solve China's environmental pollution problems.[19]

Halting as they may have been, these bureaucratic endeavors were put on the back burner after Mao's death and the transition to a market economy under his ultimate successor Deng Xiaoping. The reform era led to efforts to seek market-oriented solutions to social problems while discouraging state responsibilities in the social policy sector, forcing health-related public institutions to give priority to making money at the expense of other activities. Rebuilding state institutions after Mao led to the establishment of new bureaucracies with narrowly defined responsibilities, groups such as the State Food and Drug Administration (SFDA), Chinese Center for Disease Control and Prevention (China CDC), and the Hygiene Supervisory and Monitoring Institutes, whose isolated mandates each fragmented the capacity of the centralized health system to respond effectively to environmental risks.[20]

This is not to say that the government did nothing. In 1982, China kicked off the National Improved Stove Program, seeking to reduce exposure to smoke from traditional cookstoves. In the coming decade, 180 million improved cookstoves were distributed to people in the countryside, a move that led to more than a 30 percent reduction in incidence of lung cancer.[21] The government also conducted the first major study of the impact of air pollution on mortality in twenty-six cities, covering the period 1976–1981. The study was never made public, however, and further research in this area was constrained by the lack of reliable data.[22]

Starting in the 1980s, the central government also embarked on policies to improve the quality of drinking water. In my hometown, the perceived linkage between water pollution and a high incidence of esophageal cancer prompted the head of the county to launch an initiative to make tap water accessible to 87 percent of the population of 260,000, then the highest coverage rate in the countryside of Jiangsu province.[23] By the end of 1990s, China had improved the access to drinking water for approximately 216 million rural people.[24] In 1990, the government kicked off the Hygienic Cities movement, which aims to create better living conditions for urban residents. Seven years later, it also launched a campaign to build national environmental-protection "model cities," which highly values air quality. Local government officials have strong incentives to meet the standards of such movements because attaining the title of "hygienic city" or environmental-protection "model city" not only improves the image of the city and helps attract external investment for economic growth but also justifies their investment in infrastructure and other vanity projects that makes them look good in performance assessment. Indeed, research shows that the environmental-protection model cities movement encouraged people to "vote with feet" and relocate to such model cities, beefing up local housing market.[25]

In 1993, China also launched a national project to eliminate iodine deficiency, which had placed some 400 million people in the country at risk for a range of disorders, including enlarged thyroid, stillbirths, and stunted growth. The iodine campaign was successful in part because it was cost-effective – fortifying salt with iodine in 94 percent of the country cost the government only about $152 million.[26]

But beyond these actions, few substantive measures were undertaken to address the country's vast environmental health problems, especially the health effects of air, water, and soil pollution.[27]

BUILDING A LEGAL AND INSTITUTIONAL FRAMEWORK. Since 1979, the state has sought to establish a legal framework to safeguard the environment and to prevent pollution, which justified the setup of standalone environmental protection agencies. In 1987, the state formulated the Air Pollution Prevention and Control Law. One year later, the State Environmental Protection Agency (SEPA), the predecessor of the State Environmental Protection Administration, achieved independent status from the Ministry of Urban and Rural Construction. Under the capable leadership of Qu Geping, SEPA soon became an important supporter of environmental health. The primary objective, as enshrined in the revised Law on Prevention and Control of Water Pollution (1996), was to impose total emission control. In 1996, China introduced the Air Pollution Index (API) to measure air quality, which was based on the concentration level of sulfur dioxide (SO_2), nitrogen dioxide (NO_2) and PM10 (particulate matter 10 micrometers or less in diameter). The systematic indicators of air quality, while questionable in terms of reliability, enabled policy-makers to have a better idea of the seriousness of air pollution in China. In light of the worsening pollution, Beijing party chief and Politburo member Jia Qinglin in 1999 spoke bluntly that "we had no exit." SEPA's head, Xie Zhenhua, echoed that view, vowing "we should not allow a seriously polluted Beijing to enter the 21st century."[28]

The building of a legal and institutional framework on environmental protection opened more space for scientists, NGOs, the media, and environmental activists to raise awareness of environmental health and to support the elevation of the issue on the government agenda. China's participation in global environmental governance talks, such as the 1992 UN Conference on Environment and Development, facilitated the rise of a nascent environmental health protection constituency by promoting a new vocabulary for environmental governance, including PM, sustainable development, public participation, and market-oriented policy tools. As early as 1995, the PM2.5 levels in four major Chinese cities were reported to be higher than the average US level. Four years later, a team of scientists from Tongji University, Hong Kong University of Science and Technology, and General Motors conducted a one-year joint study on PM2.5 concentration in Shanghai and found the city's air pollution level to be extremely high even based on the relatively lax government-set standards.[29]

ELEVATION OF THE ENVIRONMENTAL HEALTH ISSUE (1998–2012)

As a result of the state's growing interest in environmental protection, SEPA was set up as a ministry-level bureau in 1998. One of SEPA's first achievements was to push for the phasing out of leaded gasoline. Initially, the ministry was not able to overcome the resistance from the biggest oil companies in China: both China National Petroleum Corp (CNPC) and the China Petroleum and Chemical Corporation (Sinopec) enjoyed the same bureaucratic rank as SEPA and were powerful enough to block it from setting gasoline quality standards. The resistance from oil companies was not broken until the State Council promulgated a regulation banning the production of leaded gasoline. In July 2000, four years after the United States banned the sale of leaded fuel for use in on-road vehicles, China achieved the goal of not producing or selling leaded fuels. The blood lead level of young Chinese (ages 0–18 years old) has gradually decreased over time.[30]

In 2000, China also revised the Law on Prevention and Control of Air Pollution. The revised law was accompanied by the introduction of a strict policy limiting the growth of air pollution from coal-fired power plants. Flue gas desulfurization (FGD) systems were installed, and their use was encouraged with the implementation of the new emission standards (GBI3223-2003), which stipulated the upper limits for air pollutant emission of these plants. By 2010, 86 percent of the coal-fired power plants in China had installed the FGD systems.[31]

Still, in the first decade of the century environmental health was not elevated to a central role in the government agenda. Driven by the "development first-environment later" mindset, Chinese leaders did not take environmental protection or public health as seriously as they indicated. Large-scale development campaigns that were likely to be environmentally degrading continued to be launched – the "Go West" campaign, for example, excluded SEPA and the Ministry of Health (MOH) from the nineteen-agency group identified at the campaign's outset.

Against this background, international organizations played a critical role in connecting the environment and health sectors. Triggered by a growing concern about the environmental and human health costs of rapid economic growth, China approached the World Bank in 2003 to develop an estimate of the costs of air and water pollution, including the pollution's impact on human health. The draft report that came out four years later confirmed huge economic costs and human health impacts linked to pollution, including an estimate of as many as 750,000 premature deaths annually.[32]

Out of fear of the social consequences of making the information public, the government asked the World Bank to remove the premature deaths estimate from the report. Yet mounting evidence from other channels made an ostrich policy impossible to sustain. According to a WikiLeaks document, a senior SEPA official admitted at a 2005 international conference that air pollution was leading to "more and more serious" health problems.[33] During 2004–2005, Chinese media's frequent coverage of the high cancer incidences in the Huai River basin in eastern China drew the attention of Premier Wen Jiabao. In March 2005, the State Council approved emergency plans to improve drinking water safety in rural areas. The government also began to provide funding to help "cancer villages" access clean drinking water.[34] At the request of Premier Wen, a joint research team consisting of experts from the MOH and SEPA was dispatched to the Huai River area in an attempt to understand the link between water pollution and high rates of cancer in the area. Headed by Yang Gonghuan, the joint team came up with a report that was submitted to the State Council in May 2006. In her briefing with Premier Wen and other central government officials in August, Yang highlighted the emergence of new cancer clusters in China and their connection to water pollution.[35]

That cooperation between health and environmental departments was driven largely by the requirements from international organizations. In 2004, China participated in the first WHO/UNEP/ADB High-level Meeting on Health and Environment in ASEAN and East Asian Countries, held in Manila. The meeting called for each member state to come up with a national environmental health action plan. In response, SEPA and MOH co-sponsored the first national environmental health forum in 2005 to discuss the framework and substance of the action plan. In June 2006, senior officials from both government agencies met to discuss how to build a long-term coordination mechanism on environmental health issues. The following year, the government released the National Environmental Health Action Plan (2007–2015), the first programmatic document guiding China's environmental health work. In addition to establishing coordination mechanisms, the action plan identified SEPA and MOH as the two lead agencies in forming a leading small group on environmental health. Furthermore, the action plan promised to conduct joint SEPA-MOH surveys and research on environmental health, build a database on environmental health monitoring, and improve the sharing and circulation of environmental health-related information.

IMPACT OF THE 2008 SUMMER OLYMPICS. China's successful bid to host the 2008 Summer Olympics highlighted environmental health issues in Beijing, one of the most polluted cities in the world, and provided formidable motivation to improve the environment in general, with a particular focus on air quality. When Beijing was competing with other cities to host the games, some International Olympic Committee (IOC) delegates expressed concerns about pollution in China, especially in Beijing, and its effects on athletes' health. In response to such concerns, China assured the IOC that it would reduce both air and water pollution in the city in advance of the games,[36] promising specifically to meet the WHO air quality guidelines. From 2001 through 2008 China invested 140 billion *yuan* (approximately $17.5 billion) and implemented more than 160 programs in controlling coal-burning pollution, vehicle emissions, industrial pollution, and dust.[37] In 2003, Beijing's Environmental Protection Bureau (EPB) began to broadcast data on the city's air quality through the local TV networks.

In spite of such efforts, a 2007 United Nations Environment Program (UNEP) assessment report noted that air pollution, as indicated by extremely high PM10 concentration levels, remained the single largest environmental and public health issue affecting Beijing.[38] This led IOC president Jacques Rogge to imply that the IOC might have to consider postponing or delaying the Olympic Games for endurance sports, such as cycling, because of the acute and chronic deleterious effects of PM on the athletes' respiratory and cardiovascular systems.[39] Interestingly, it was during this period that Chinese leaders began to show serious concern about the effects of air pollution on their own health. In the same year, powerful air purifiers manufactured by the Broad Group – each cost about $1,300 – were installed in the Zhongnanhai leadership compound in central Beijing to "protect state leaders' breathing health."[40] Central leaders got a glimpse of the seriousness of air pollution in the city when seeing "ink-like dirty water" come out when cleaning the soot-laden filters.[41]

In response to the IOC's warnings, the government turned to extraordinary measures: it shut down or relocated factories in Beijing and surrounding provinces, halted major construction projects, and took cars off the road on alternate days.

Beijing's efforts to increase the number of the so-called blue sky days (i.e., when emissions fall below official targets) drew criticisms that environmental officials were also manipulating pollution data to meet the government targets.[42] The fear of adverse press coverage on air pollution in Beijing led to efforts to downplay pollution problems in other major Chinese cities. In June 2008, a top Chinese lung expert, Zhong Nanshan, said that in the southern city of Guangzhou, people's lungs were turning black after their fifties.[43] His remarks fell on deaf ears until March 2013.[44]

In the absence of public discussion of air pollution before and during the Olympics, most Chinese people had no idea how bad air pollution was in the country, nor did they know what smog or PM was. They were shocked to see US track cyclists arriving in Beijing for the Games in early August wearing facial masks. Many perceived this occurrence to be an insult to China. An opinion piece on the website of the official *People's Daily* called it a "dirty trick" played by the "politicians and anti-China forces of some Western countries."[45] Lambasted by the Chinese media, the four US cyclists apologized.[46]

Many Chinese government officials were simply in denial about poor air quality in Beijing. When asked about whether athletes should bring facial masks to China, the deputy head of Beijing EPB said: "I don't think it is necessary to consider wearing masks. If you really want to wear them, it just adds more weight to your luggage. I don't think it is needed."[47] (Thanks to the government ad hoc measures, Beijing managed to achieve relatively clear skies during the Games; the air quality went back to normal shortly afterwards.)

US EMBASSY'S MONITORING OF AIR QUALITY. The Beijing Olympics also drew attention to the US Embassy's monitoring of air quality and its sharing of the information with its staff and American expatriates. Since 2003, API numbers issued by the Beijing EPB were around 99 for most days. Any API less than 100 classified the day as a "blue sky" day, which, given the gray skies overhead, led one US Embassy official to question the credibility of the official data.[48] In February 2008, the US Embassy installed a rooftop PM monitor on top of the mailroom building to measure PM2.5 concentration in the Embassy compound area. Five months later, the device was programmed to post hourly air quality index (AQI) numbers to an embassy-managed Twitter account (http://twitter.com/beijingair). According to embassy officials, the original idea behind this initiative was to inform staff about the air quality around their compound, but a "no double standards" requirement prompted the Twitter feed so that all American citizens traveling and residing in Beijing had access to the data.[49]

The data posted by the embassy, and the increasing popularity of its Twitter account, quickly laid bare the shortcomings of Beijing's air quality reporting system. First, the embassy recorded levels of PM2.5 and published AQI figures according to the US Environmental Protection Agency (EPA) standards. Even though the Beijing EPB was already measuring PM2.5 at the time,[50] the government only released public data on PM10 (larger particles considered less dangerous to human health than PM2.5), SO_2, and NO_2, calculated according to the government's own API. Second, instead of posting real-time air quality data on an hourly basis, the Chinese government only published one daily average online at noon to describe the previous 24-hour period. Posted after the fact, the Chinese data offered little value to people who wanted guidance on whether or not it was safe to engage in outdoor activities on any given day.[51] As Ma Jun, a leader of an environmental NGO that tracks pollution data in China, later recalled:

> Before the [PM2.5-based] index was public, even on smoggy days, my child ran around outdoors at school. I worried because I know how bad it was, but other parents didn't know.[52]

While the United States did not actively promote the availability of air quality data and insisted that the monitor was only a resource for the health of the embassy community, its Twitter site soon became known to some Chinese residents as well. Often painting a bleaker picture than the official Chinese pronouncements, the embassy air quality data more accurately reflected the actual experiences of Beijing's residents. After *Time* magazine published an online story about the embassy's air monitor on June 19, 2009,[53] the number of registered followers of the Twitter account increased significantly, from approximately 400 to more than 2,500, with at least three quarters of the new followers being Chinese. The actual number of people viewing the data was probably larger, considering that the site was public to non-followers as well. Local and international press coverage spiked after that, setting media outlets such as the *South China Morning Post, China Daily*, and Sina.com ablaze with Chinese "netizens" commenting on the issue. By June 2012, there were more than 19,000 registered followers of the embassy's Twitter handle.[54] By summer 2015, it had attracted one million hits a month.[55]

Irritated by the growing popularity of the embassy's air monitoring practice, China's Ministry of Foreign Affairs (MFA) privately lodged protests against the US Embassy in July 2009. One Chinese diplomat argued that Beijing EPB should be the sole authoritative voice for public announcements on the city's air quality. The characterization of Beijing's air quality as "unhealthy" or "very unhealthy," he said, diminished the government efforts to improve air quality. He also pointed to the use of US standards for measuring air pollution as "confusing" to the public and "insulting" to the Chinese government – an action that could lead to "social consequences." Despite these objections, US Embassy officials stood their ground, remaining firm that they would not consider discontinuing its monitoring program until Beijing began to publish PM2.5 on a real-time basis.[56]

The Chinese environmental protection agencies had the technical capacity to measure PM2.5 levels, but they were reluctant to make them public. An anonymous MEP official commented, "If we were required to implement PM2.5 air quality standards, many places would exceed the standards by a large margin."[57] The implication was that if the Chinese government were to fully disclose PM2.5 information, it would make environmental protection departments look bad at policy implementation. Zhang Yuanhang, a leading expert in environmental science in China, concurred, saying that the percentage of cities meeting the air quality standards would drop from 80 percent to 20 percent if PM2.5 was incorporated into the national monitoring system.[58] Apart from the bureaucratic incentive problem, reporting on PM2.5 was considered a sensitive political issue in China. Rather than implementing a new and more effective monitoring system, China officially blocked Twitter in 2009. Until 2011, government meteorologists still denied emissions as a cause of smog. In November that year, Nanjing City Bureau of Meteorology released data on local PM2.5 through its account on Weibo, a Chinese micro-blogging platform similar to Twitter. The data were removed immediately and, reportedly, the person who issued the data was disciplined.[59]

The government resistance did not stop locals from gaining access to the embassy's Twitter service using VPN technology. In late October 2010, celebrity real estate developer and opinion leader Pan Shiyi forwarded to Weibo a screenshot of the US Embassy Twitter account showing the PM2.5 level was 387 ("Hazardous"), with the comment "Oh my God, the air is toxic!" Within hours, Pan's message was forwarded more than 7,000 times. The embassy's air quality data were in sharp contrast to Beijing EPB's website that showed that the air was "slightly polluted."[60] The gap led people to suspect the government was not telling the truth. A survey conducted by the *China Youth Daily* in early November found that nearly 70 percent of the respondents said that the government data were not compatible with what they actually felt.[61] On the morning of November 17, 2010, some Beijingers, who were used to the frequent "bad" and "hazardous" readings on the US Embassy Twitter site, woke up to find themselves appalled by the levels of PM2.5, which had surged past 500 – about twenty times higher than the level deemed acceptable by the WHO. Triggered by the reading, which was far off the normal scale, the automatic system labeled the day's air "crazy bad," a stark, dark-humor analysis embedded into the embassy monitor's program. While embassy officials quickly deleted this display and replaced it with the term "beyond index," the original tweet already had reached a large audience who found the description shocking but significant.[62]

DEMOCRATIZATION OF AIR QUALITY DATA. These events exposed the crudeness of Beijing's air quality monitoring system, generating growing pressure on the Chinese government to acknowledge the scale of the problem and to take proactive measures. At the Chinese People's Political Consultative Conference (CPPCC), held in March 2011, the Central Committee of China Association for Promoting Democracy, a minor political party that is part of the CCP's United Front, submitted a proposal pleading for routine PM2.5 monitoring in China's air quality monitoring system.[63] That November, Dr. Zhong Nanshan asked the government to publicize PM2.5 data, claiming "dust haze" was more dangerous than SARS. Zhong's petition resonated. An online poll conducted by Pan Shiyi on Sina Weibo found that 92 percent of the 40,000 respondents agreed that "the authorities should adopt PM2.5 standards this year."[64] Frustrated that the government was reluctant to introduce the new standard, Chinese netizens launched a movement called "I gauge air quality for my motherland." Apparently inspired by the popular 1960s propaganda poster motto "I exploit crude oil for my motherland," the slogan stoked a grassroots campaign encouraging the public to take air quality monitoring into their own hands.[65]

The rising public awareness forced the government to relent and be more forthcoming with health-related environmental information. In January 2012, Beijing became the first city in China to publicize PM2.5 data. Zhejiang province and Shanghai followed suit and revealed plans to release PM2.5 data for the first half of the year.[66] On February 29, Premier Wen Jiabao presided over a State Council executive meeting to approve the new ambient air quality standards (GB 3095–2012), which asked all major Chinese cities to conduct PM2.5 monitoring. By the beginning of 2013, the system was ready to publish PM2.5-based AQI (instead of API) in real-time in seventy-four cities throughout the country, making the worsening air pollution quantifiable, observable, and undeniable. By the end of 2017, China had built a nationwide air quality monitoring network, with nearly 1,500 MEP-controlled monitoring stations covering 388 cities at or above prefectural level. In addition, there are 3,500 monitoring stations managed by provincial or city governments.[67]

When the US Embassy began tweeting PM2.5 data, few could have anticipated its impact on China's environmental health policy. By tapping into the transformative power of democratized data, the US mission in China unintentionally made the Chinese government more transparent in sharing environmental health information. The shift to a new monitoring system, in turn, highlighted the seriousness of China's environmental health problem and forced the government to take more action on pollution. Shanghai's three-year action plan for ambient environmental protection, for example, made the prevention and control of PM2.5 a policy priority.[68]

Prior to the introduction of AQI in China, manipulation of air quality data was very common at the local level. An analysis of data on daily air pollution concentrations over the period 2001–2010 found that about half of Chinese cities reported fake PM10 levels, which led to a discontinuity at the cut-off point 100 for "blue-sky days."[69] The nationwide installation of monitoring stations that can automatically generate air quality data thus helped mitigate the data reliability problem.[70] Furthermore, because PM2.5 are representative pollutants of compound air pollution, the focus on their concentration enabled government officials to gain a relatively comprehensive understanding of the local industrial and polluting landscape. As Premier Li Keqiang later admitted, PM2.5 was not just a reminder to the public about taking protective measures against pollution; it also puts pressure on the government to act.[71] The democratization of air quality data helps explain why air pollution prevention and control became the top priority in addressing environmental health hazards. As a senior Chinese editor said in his Wechat friend circle years later, "in my scope of imagination, [smog] was the only thing the increasingly authoritarian government cannot cover up."

DECLARING WAR ON POLLUTION (2013 – 2017)

Premier Wen's decision to publicize PM2.5 data occurred when the party-state was plunged into a larger credibility crisis. On February 6, 2012, Wang Lijun, former police chief of Chongqing, fled to the US Consulate in Chengdu to seek asylum. The incident exposed the murder of British businessman Neil Heywood by Gu Kailai, the wife of Chongqing's party chief, Bo Xilai. Coincidently, the paranoia that led Gu to murder was, in part, triggered by heavy metal poisoning when she was taking a herbal remedy.[72] The incident blew open China's biggest political scandal in decades while setting in motion a cascade of events leading to the downfall of Bo, a rising political star who coveted a position on the all-powerful Politburo Standing Committee. Two months later, Chen Guangcheng, a blind Chinese rights lawyer, evaded house arrest in his village and sought asylum in the US Embassy in Beijing. Both scandals dealt major embarrassments to the Chinese government. The capital's newfound motivation to clean up its air lost its coherence as it became entangled in domestic politics and foreign policy challenges. In an effort to prevent any more unscripted events from interfering with the leadership succession at the impending 18th Party Congress (scheduled for November 2012), the Chinese government asked the US Embassy to stop releasing Beijing's air quality data. Contending that the monitoring of air quality was a sovereign power of government in the host country, at a press conference on June 5, 2012, a vice minister of the MEP implicitly accused the United States of illegally interfering in China's domestic affairs.[73]

Alleging that a diplomatic mission's readings of local air pollution level were illegal only invited more ridicule from the increasingly skeptical Chinese netizens.[74] "Our great nation should go tit-for-tat and issue PM2.5 readings from our own embassies, make our embassies the safest place for the Occupy Wall Street people, and make our country proud," mocked one Chinese commentator. "Do you think people are blind? How many blue-sky days has Beijing had lately? Do you think ordinary people will only believe your own statements?" grumbled another.[75] As Susan Shirk and Steven Oliver noted in a blog post: "This self-defeating action (from the government) is symptomatic of a panicky leadership with a severe credibility problem."[76] Amidst speculation of a leadership succession, Xi Jinping, the anointed party leader, disappeared from public view for two weeks, and a scheduled meeting with Hillary Clinton, the US Secretary of State, was canceled in September.

XI JINPING'S "TWO MOUNTAINS" THESIS. Ultimately, the dust settled at the 18th Party Congress, which formally appointed Xi Jinping China's new political leader. Xi's designation opened a political window for a shift in China's environmental health policy. Xi had learned about the importance of balancing economic development with environmental protection when he was the party secretary of Zhejiang province (2002–2007). Before serving in Zhejiang, Xi reportedly already had been exposed to important development concepts, such as sustainable development, green GDP, and the human development index (HDI).[77] He updated his thinking on development after a series of mass incidents in the province in 2005. Indeed, the first environmental health-related mass incident in China arose in Dongyang in Zhejiang, involving over 20,000 villagers and leading to thirty-three injuries. This was followed by two other widely reported violent conflicts with police in Xinchang and Changxing, both also in Zhejiang province. All three incidents, as a Chinese scholar observed, broke out when local residents found themselves beset by a pollution-induced crisis. In Changxing, for example, rapid expansion of the battery industry in the county was associated with a large number of cases of child lead poisoning.[78] Drawing lessons from these events, Xi came up with his "two mountains" thesis in August 2005:

> Make sure not to repeat the old development path or be obsessed with the old path ... In the past we talked about the need to have clear water and green mountains as well as mountains of gold and silver. However, in fact clear water and green mountains are as good as mountains of gold and silver.[79]

After that, Xi used a pen name, *Zhexin,* to publish an opinion article in the official *Zhejiang Daily* to elaborate on the relationship between environmental protection and development. He argued that a clean environment could produce wealth if Zhejiang translated its comparative ecological advantages into advances in agricultural, industrial, and tourism development.[80] The administrative turnover in November 2012 thus produced a key leader who was ready to reset the agenda on environmental health.

THE 2013 PM2.5 CRISIS. Still, it was the PM2.5 crisis of 2013 that weaved the problem, policy, and political streams together, pushing all toward serious change. On the first day of the year, Beijing and surrounding regions were hit with astonishingly high levels of air pollution. On the evening of January 12, in what *The Economist* magazine called the

"Blackest Day,"[81] AQI readings in Beijing went off the charts – with a reading of 755, it was nearly twice the level the US EPA deems "hazardous" and thirty times the value set by the WHO as acceptable quality (25 micrograms per cubic meter).

Smog was suddenly omnipresent in China. Indeed, the entire country experienced the worst haze outbreak since 1961, when systematic meteorological data first were available. With the seasonal restart of northeast China's coal-powered heating system in late October, PM densities reached record highs in that region as well. In Harbin, the capital city of Heilongjiang province, PM2.5 concentration rose to 1,000 *ug*/m^3, a level even higher than Beijing's worst level. In December, a heavy smog engulfed 25 of China's 31 provinces and municipalities, covering over 100 large or medium cities, spanning over 1.4 million square kilometers, and affecting over 800 million people in China.[82] Even Sanya, the city in Hainan that allegedly has the best air quality in China, and Lhasa, the capital city of remote Tibet, whose residents proclaim it has almost perfect air quality, were not spared. Not only did the smog stretch out geographically, it also lasted longer than ever before. The nation's average number of smog days in 2013 was 29.9, 10.3 more days than in 2012.[83] The number of smog days in heavily industrialized regions around Beijing, Tianjin, and Hebei, and in the Pearl and Yangtze River Delta regions was more than 100, and in some cities, higher than 200.[84]

The unprecedented smog outbreak contributed to soaring morbidity and mortality rates. The number of children treated at Beijing Children's Hospital for respiratory ailments reached 7,000 a day in January – a five-year high. In Harbin, hospital admissions surged by 30 percent in October.[85] The same month, the WHO's specialized cancer agency, the International Agency for Research on Cancer, announced that sufficient evidence shows that exposure to ambient air pollution and PM causes lung cancer.[86] Chinese scientists found that exposure to PM2.5 in January alone was associated with 1,416 premature deaths in four major cities (Beijing, Shanghai, Guangzhou, and Xi'an).[87] The direct economic cost of smog on transportation and health in the month was estimated to be at least 23 billion *yuan* ($3.7 billion).[88] The death toll soared to 25,700 for the year, a mortality rate, if extrapolated to all thirty-one provincial capitals, higher than that caused by smoking or traffic accidents.[89]

The 2013 smog sparked a national outcry while drawing international media attention. "PM2.5" as a term received three million mentions on Sina Weibo in January 2013, compared to the mere 200 mentions it drew two years before.[90] Dissatisfaction with government inaction was evidenced in the voting results in electing government ministers at the National People's Congress (NPC) meeting in March 2013. Minister of Environmental Protection Zhou Shengxian received the lowest votes among the twenty-five ministers.[91] In April, the National Cancer Prevention and Care Week released a slogan "Protect the Environment, Keep Cancer Away." As a *New York Times* article observed, "[t]hat even the Chinese state is highlighting a link between the poor environment and cancer reflects an atmosphere of deep concern, verging on panic, over public health, as more and more people ask: Is China killing itself in the pursuit of spectacularly fast, very dirty, economic growth?"[92]

The health, economic, and social implications of the "airpocalypse" of early 2013 prompted the government to act in a more timely manner. In contrast to previous environmental protection campaigns, this time the focus was squarely on health. Particulate matter became the target of government intervention. In September, the State Council unveiled the five-year "Action Plan for the Prevention and Control of Air Pollution" (*qi shi tiao,* or "ten measures for air"), which aimed (1) to curb PM10 level in cities above the prefecture level by at least 10 percent against the 2012 level; (2) to increase the days with good air quality year on year; and (3) to cut the PM2.5 level in Beijing-Tianjin-Hebei, and the Yangtze and Pearl River deltas by 25 percent, 20 percent, and 15 percent, respectively. The annual PM2.5 concentration in Beijing was to be kept at or below 60 microgram per cubic meter (ug/m^3).

This was the first time in the history of the People's Republic of China that an environmental action plan was introduced in the name of the State Council.[93] The move signaled the shift of air pollution policy from an item on the "specialized agenda," which referred to the agenda of government agencies, to the "general agenda," where the top leaders become actively involved.[94] Under the new policy, local governments were asked to implement tougher controls on pollution, industrial production, and coal consumption. In response, the Beijing city government released a five-year plan that promised, among other items, to cut coal burning by more than 50 percent. Later that year, China pledged 3.4 trillion *yuan* ($549 billion) in the 12th Five Year Plan (2012–2017) for cleaning the air and implementing PM2.5-defined pollution targets for major cities. Concerns about the health consequences of China's growing environmental woes led to increased funding for environmental health research. With funding from the National Natural Science Foundation of China, the number of studies on how pollution affects fertility tripled from 23 in 2008 to 68 in 2013.[95] In March 2014, Premier Li Keqiang declared "war on pollution," vowing to give harsher punishments to polluters and to discipline officials neglecting their environmental duties. He also indicated a willingness to involve non-state actors in the process by asking "the whole society, including the government, business, and citizens" to work together in the fight against pollution.[96] In the same month, NPC updated the country's environmental protection law for the first time in twenty-five years, specifying environmental health as an important component of environmental protection while approving stricter punishments for factories that pollute.

Top Chinese leaders offered clear support for these measures. In 2014, seven Politburo Standing Committee members made 560 written instructions on environmental protection, including seventeen from Xi Jinping, seventy-three from Li Keqiang, and 456 from Zhang Gaoli (whose portfolio in the Committee spanned the fields of economic development, natural resources, and the environment).[97] During the Asia-Pacific Economic Cooperation (APEC) summit in November, 2014, Xi admitted that smog was his top priority: "These days, the first thing I have been doing after getting up is to check how Beijing's air quality is and hope there is less smog so that all these guests from distant places will feel more comfortable."[98] China's environmental health crisis also softened China's attitude toward global carbon emission control. During the APEC summit, President Xi and US President Barack Obama signed a joint agreement in which China promised to curb the growth of greenhouse-gas emissions by 2030.

A NEW POLICY FRAMEWORK. At the top level, there were changes afoot that suggested the willingness of the Party to defuse public demands for better environmental health. In April 2015 the central party leadership restated its plan to implement "ecological civilization reforms" and emphasized the need to reconcile tensions between economic development and the environment. The same month, the State Council unveiled the "Action Plan for the Prevention and Control of Water Pollution" (*shui shi tiao*, or "ten measures for water") that vowed not only to shut down factories that pollute water supplies, but also to increase the share of water suitable for human consumption in seven major river basins to more than 70 percent by 2020. This was followed by the release of the "Action Plan for the Prevention and Control of Soil Pollution" (*tu shi tiao*, or "ten measures for soil") one year later, which aimed to ensure 90 percent of currently polluted farmland was usable by 2020. So, by April 2015, the government had put in place three action plans to tackle the country's air, water and soil pollution problems. The action plans joined the Water Pollution and Control Law (2008), the revised Environmental Protection Law (2014), the updated Food and Safety Law (2015), the amended Air Pollution Prevention and Control Law (2015), and have since been joined by the Nuclear Safety Law (2017), and the Soil Pollution and Control Law (2018). Together, these set the new policy framework governing environmental health issues.

Over time, the environmental protection policy goals have become increasingly geared toward mitigating pollution's health effects. Rather than solely emphasizing capping the total emission level, as enshrined in the 11th and 12th Five-Year Plans, the 13th Five-Year Plan (2016–2020) acknowledged the dangers posed by pollution and for the first time proposed specific targets for reducing PM2.5. The growing emphasis on health found its resonance in Xi Jinping's speech at the National Health Conference in August 2016. In the speech, President Xi called for full protection of the people's health, stressing that "health is a prerequisite for people's all-round development and a precondition for economic and social development."[99] Furthermore, Xi linked the environment to health by saying that a sound environment is the "cornerstone" of the people's lives and health. In tandem with Xi's speech,

China issued a "Healthy China 2030" plan, seeking to provide comprehensive healthcare for all by 2030, to improve average life expectancy to seventy-nine years and match rich-country levels in health outcomes. In June 2017, environmental health standards were incorporated formally into the state environmental protection regime.

The 19th Party Congress, held in October 2017, gave additional impetus to pollution control and environmental health. In Xi Jinping's political report delivered to the Congress, the word "environment" was mentioned eighty-nine times, compared with only seventy uses of the word "economy" – a reversal of the emphasis placed on those words at the 18th Party Congress five years before.[100] The 2017 report has a full chapter on the environment, entitled "Speeding up reform of the system for developing an ecological civilization and building a beautiful China." In view of the apparent loss of momentum on reform on other fronts – indeed, no other chapter has "speeding up reform" in its title – ecological reform has risen as a top policy priority, while "ecological civilization" now has become part of Xi's official governance ideology. Meanwhile, the term "beautiful China" ties the environment to a sense of national pride and China's international ascendance. This "environment-in-all" approach, which seeks to integrate environmental considerations into the economy, politics, culture, and society, represents an important change in CCP's ideological foundations and has profound and positive implications for environmental health in China.

LIMITATIONS OF ENVIRONMENTAL HEALTH POLICYMAKING

In characterizing China's policy-making process, two Chinese scholars have proposed a "truncated" model, in which key political leaders have the flexibility to make impromptu decisions, but these decisions tend to be based more on political instinct than on calculated rational analysis conducted by technical experts or bureaucrats.[101] That is exactly the case here. While central leaders have demonstrated strong political commitment to pollution control, the policy shift is neither thorough nor complete.

The following gaps and deficiencies can be identified in China's environmental-health policymaking: the lack of convergence of science and policy, asymmetry in the information available to the state and to common members of society, delays in creating a health-oriented environmental policy, and a continuing conflict between development and environmental health. These shortcomings all raise further questions about whether the government policy response can be translated into sufficient improvements on the ground:

LACK OF CONVERGENCE OF SCIENCE AND POLICY. Collaboration between researchers and policymakers leads to more effective public policies.[102] Historically, however, the reactive, crisis-driven policy process in China fostered decision-making based on inadequate research or evidence. As one Chinese political scientist observed, "our policy often reflects the will of leaders. It can be introduced very quickly. Once it is triggered and meets up with the public ethical orientation, there will be little query or scientific verification."[103] A leading Chinese health scholar, Meng Qingyue, concurs, noting that while the context of policy-making has changed significantly in the post-Mao era, "many policies are still formulated on the basis of weak or no evidence."[104]

As suggested by some Chinese scientists, China's failure to conduct adequate research about its own air pollution problem explains why it had to transplant AQI values directly from the United States. The problem is that the Western smog forecast model may underestimate the seriousness of China's air pollution.[105] When Chinese government hurriedly devised the national action plan on air pollution in 2013, for example, it set targets based on incomplete scientific data. At that time China had not yet established the National Air Quality Monitoring Network, and Chinese scientists were still debating the causes of and contributors to smog in China. Premier Li Keqiang acknowledged that China's air-pollution measures have focused on limiting coal burning, car emissions, and flying dust without tackling other pollutants, such as ammonia released by nitrogen fertilizers used in agriculture, which may contribute up to 20 percent of the smog in China.[106] The same problem also occurred in soil pollution control. The pollution reduction targets in the action plan on soil pollution were set without a clear understanding of the size and distribution of the polluted farmland.[107]

Moreover, the pollution control targets were set without the benefit of adequate research into the impact of pollution on human health. In order to have a thorough understanding of the health effects of pollution, long-term cohort studies – a type of medical research that looks at groups of people to investigate the causes of disease and to establish links between risk factors and health outcomes – are essential. By the time China promulgated its new air quality standards, China had conducted only one retrospective cohort study on the health effects of air pollution, and its usefulness was questioned due to the lack of data on PM2.5.[108] It was not until five years later that Chinese scientists published results of the first prospective cohort study of the association between long-term exposure to PM2.5 and nonaccidental and cause-specific mortality in China.[109]

Similarly, the Beijing city government did not begin to conduct a cohort observation on the relationship between smog and lung cancer until December 2016, and only in April 2017 did China launch a major study focusing on the causes, health impacts, and governance issues associated with heavy air pollution.[110] A senior health official in January 2017 admitted "at present, our country does not have conclusive clinical studies and related statistical data about the connection between smog and cancer, and it is still too early to draw conclusions on smog's harm to health, especially its long-term health effect."[111] As a result, some of the pollution-control targets seem arbitrary: How did policymakers decide that the level of PM10 should be brought down by 10 percent nationwide? Is a nationwide goal, as opposed to specific targets for the worst regions, even appropriate?

And sometimes the targets were far from enough: Beijing was asked to bring its PM2.5 level below 60 ug/m^3, yet the WHO recommends a maximum annual mean of 10 ug/m^3. The lack of an evidence-based, science-backed policy process also was indicated by some of the buzzwords used by the government officials in tackling pollution. A US environmental health expert told me that it was difficult to convince senior provincial government officials that environmental pollution cannot be "eliminated."[112]

The inability of China to set up underlying databases on the health effects of pollution can be attributed to the lack of a stable, long-term environmental health funding mechanism. The National Natural Science Foundation of China (NSFC), for example, only supports research projects for three to four years, yet prospective cohort studies usually require at least ten-year follow-up periods.[113] Bureaucratic fragmentation makes the situation even worse. An MEP official was blunt:

> Different departments were assigned projects worth several million to 10 million *yuan*, but the studies were not systematic. They were fragmented. It was not until they started to make environmental quality standards that they found lots of data was not available. Since [China] does not have its own data, it has to use data from other countries.[114]

Political considerations also have limited the space for Chinese scientists to conduct needed research that would otherwise inform the decision makers and the public in the environmental health policy process. According to Ma Jun, director of the Institute of Public and Environmental Affairs (IPE), access to even the limited research on the relationship between air pollution and lung cancer has been restricted because the topic itself is politically sensitive: "The government does not want to cause panic among the public."[115]

Meanwhile, fake or inaccurate data continue to be a challenge for environmental health-related policymaking. Based on government data, a study on the cost effectiveness of emission control measures concluded that northeastern provinces did the best among all Chinese provinces in addressing water pollution in 2014.[116] This study proved useless as soon as the MEP found that thousands of polluters in Northern China had faked emission data.[117] Existing test- and measurement-related regulations (drafted with the help of manufacturers of test and measurement equipment) also have constrained China's ability to introduce new measurement and monitoring technologies for environmental governance, causing accuracy problems in collected pollution data. Because of fake and/or inaccurate data, many Chinese scientists argue that government policy on pollution control falls short of technical support or rationale.[118]

STATE-SOCIETY INFORMATION ASYMMETRY. In December 2016, the Council on Foreign Relations hosted a workshop in New York on

China's environmental health. Workshop participants identified public engagement as a critical factor in pushing for real change in environmental health-related policy. International experience suggests that cities where government officials report stronger pressures from their citizens to improve the environment led to better environmental performance, and government involvement is essential to fostering grassroots mobilization.[119]

These findings have profound policy implications for China, which still seeks to suppress social movements demanding improvements in environmental health. Despite official policy rhetoric and progress being made on transparency, the information flow regarding environmental health issues continues to be turned on or off like a faucet by the government. Water quality data released by the MEP only cover 113 major cities in China. Everybody knows water quality in China is bad, an anonymous scholar remarked, but how bad it is remains a state secret.[120] Similarly, soil pollution in China has become an extremely sensitive topic and the state has prohibited public access to the relevant survey data. In 2012, a directive issued by Shaanxi provincial EPB warned that those who divulge soil pollution survey data would be severely punished. One year later, the MEP rejected a request from a Beijing lawyer who applied to publicize information on the first-ever nationwide soil pollution survey on grounds that the data were a state secret.[121] He might feel better if he learned that even former SEPA head Qu Geping did not have access to the data.[122] Eight years later the government finally publicized the results of the 2010 survey, but it failed to release the raw data or full survey results to the public. Moreover, it refused to explain exactly how the survey was conducted, which would be a critical factor affecting the results.[123]

The strict information control is not so much an indication of strong state autonomy as an example of how the government environmental health policy can be hijacked by vested interests. As with all public policy issues, proactive government response to the environmental health crisis creates both winners and losers. Government agencies and nongovernmental actors promoting environmental protection would see their power and prestige increase, while polluting industries (including those who work for them) stand to lose. As Qu Geping explained, it was difficult to publicize soil pollution data because these implicate "too many departmental and local government interests."

Not surprisingly, when the film *Under the Dome* went viral, petrochemical industry insiders disputed Chai's claims that lax quality standards for petroleum were a primary contributor to China's worsening air pollution. The tension between environmental protection organizations – both official and nongovernmental – and powerful vested interests reflects an increasingly polarized Chinese society, which makes a serious and constructive discussion of the issue next to impossible. A large number of Chinese netizens debated the relationship between business practices and pollution and the role of the state vis-à-vis the market. A few Chinese officials and intellectuals began to question Chai's motives and integrity in making the film. And even one Chinese environmental expert I met in Beijing told me that the documentary contained many "illogical hints."[124] In the absence of meaningful social participation, the evolution of environmental health policy will be less effective.

DELAYS IN CREATING A HEALTH-ORIENTED ENVIRONMENTAL POLICY. Air, water, and soil pollution in China still is framed primarily as an environmental rather than a health problem. Take the ambient air quality standards. As in the United States, AQI in China is divided into six categories, but instead of having each category correspond to a different level of health concern (e.g., "unhealthy," "very unhealthy," "hazardous"), China's AQI has each category correspond to a different level of pollution (e.g., "moderately polluted," "heavily polluted," "severely polluted"). The latter provides little guidance for ordinary people to understand what the local air quality means for their health.

While the post-2012 standards for the first time set limits on PM2.5 and ozone (O_3), in many respects they are no different from the old standards in their indifference to public health. As Table 3.1 shows, the AQI 51–100 is "moderate" in the United States but "good" in China, and the PM2.5 average concentration for the same standard is 12.5–35.4 in the United States but 35–75 in China. The AQI of 101–150 is "unhealthy for sensitive groups" in the United States but considered "lightly polluted" in China.[125] The same readings that the Chinese government

TABLE 3.1 *Designations for air quality in China and the United States*

AQI value	24-hr average PM2.5 concentration ($\mu g/m^3$)		AQI category	
	China	US	China	US
0-50	0-35	0-12.4	Excellent	Good
51-100	36-75	12.5-35.4	Good	Moderate
101-150	76-115	35.5-55.4	Lightly polluted	Unhealthy for sensitive groups
151-200	116-150	55.5-150.4	Moderately polluted	Unhealthy
201-300	151-250	150.5-250.4	Heavily polluted	Very unhealthy
301-500	251-500	250.5-500.4	Severely polluted	Hazardous

Source: Author's Database

classifies as "lightly polluted" air are in the danger zone by US standards. A senior MEP official admitted that there is still a long way to go before China's new standards become comparable to those of the United States in terms of the thresholds for health-harmful pollution levels.[126] In the absence of a more health-centric AQI, not only is the practicality of the standards questionable but also the government's legitimacy will be undermined. As an environmental expert explained:

> Although most Chinese cities now have started to publish PM2.5 figures' – a major step forward – they remain evasive about the health implications of that data. The public don't understand what a daily average PM2.5 figure of 35 ug/m^3 or 75ug/m^3 means for their health. They just want to know if their elderly parents can go out for a stroll or their kids can go out to play, but the raw statistics they are given don't tell them that. In the absence of more "human" data, it is hardly surprising that so many citizens, concerned about their families, turned to the U.S. Embassy's feed and its depressing litany of warnings – exaggerations that have worsened the fear and mistrust of the government.[127]

The lack of adequate concern for public health also is evidenced in the application of two classes of limit values. Unlike the United States, which applies primary standards to public health protection, including protecting the health of susceptible populations (e.g., asthmatics, children, and the elderly), China applies Class 1 standards to national parks and nature reserves and applies Class 2 (which is less strict than Class 1) standards to urban and industrial areas.[128] The lack of a health-centric environmental policy also explains why genuine environmental health legislation remains in its "initial exploration stage."[129] It was not until January 2018 that the MEP integrated the concept of public health protection into environmental protection policy by issuing a directive on "The Trial Approach to Environmental Health in Environmental Protection."[130]

THE DILEMMA OF DEVELOPMENT AND ENVIRONMENTAL HEALTH Finally, a proactive environmental health agenda is unlikely to be met with a warm reception due to the strong development-first mentality in the policy community. China has been governed by an implicit social contract since 1989 – the regime commits to improving the material well-being of the people in exchange for the latter's support of the party's monopoly over political power.

While building an "ecological civilization" has become a "basic state policy," Xi also proposed economic targets even more ambitious than Deng in his political report of the 19th Party Congress. For this reason, GDP growth remains the centerpiece of China's policy agenda and government officials will continue to pursue rapid economic development, even if it is detrimental to environmental protection. Herein lies the quandary for Chinese leaders: the party has strong incentives to sustain robust economic growth to ensure employment and improved standards of living yet policy measures to combat pollution and improve environmental health are likely to undermine efforts to sustain GDP growth, at least in the short run.

Reports and discussions have highlighted the trade-off between China's anti-pollution measures and economic growth. The Paris-based banking and financial services firm Société Générale estimated in October 2017 that China's efforts to cut pollution could shave economic growth by 0.25 percentage points in the first half of 2018 while stoking more inflation at the factory level.[131] It's true that many Western nations only adopted serious environmental protection after achieving a relatively high level of economic well-being. A number of Chinese environmental scholars and opinion leaders, convinced by the environmental Kuznets hypothesis that high income and economic growth ultimately lead to environmental improvements, insist that economic development is a sine qua non for solving the country's environmental problems.[132] "Historical experience of developed countries demonstrates that environmental problems ultimately need to be solved by (economic) development," a leading Chinese environmental scholar wrote in the CCP mouthpiece *People's Daily*.[133] This view resonates among many other environmental experts.[134]

Justin Yifu Lin, a renowned economist in China, even argued in January 2017 that economic growth was not the primary factor behind China's smog problem, and that China should attempt to speed up growth to fundamentally solve its environmental crisis.[135] But such views fail to take account that the nature of growth needs to change for it to be sustainable over time and lead to overall gains in social and economic welfare.

Concerns also have been expressed within China that a rapid shift toward environmental protection may "undermine China's comprehensive power, inviting other countries to take advantage of us, even invade us."[136] As a result, local implementers may well be asked to fulfill a whole range of sociopolitical, environmental, and development objectives, from protecting the environment to alleviating poverty. Such centrally imposed complexity is a recipe for timidity and inaction.[137] How exactly will this dilemma drive local officials' incentives in policy implementation? This will be the subject of Chapter 4.

CHAPTER 4

Implementing Environmental Health Policy

IN JANUARY 2014, AFTER A WEEK OF PARTICULARLY HORRENDOUS AIR pollution, Beijing mayor Wang Anshun told the members of the municipal People's Political Consultative Conference that the city faced a "life-or-death" environmental crisis.[1] He probably had in mind his future career as well. Earlier, he had signed a "responsibility pledge" (*zeren zhuang*) with central government leaders and was warned that if he did not lower the city's concentration of fine particles known as PM2.5 to 60 microgram per cubic meter ($\mu g/m^3$) by the end of 2017, he would find his "head on a platter."[2]

Wang explained later that the threat from his boss was just a joke, but he could not deny the growing pressure from above to hit pollution control targets. Now that environmental health issues have been elevated to the top of the government agenda, the crucial issue is translating policy into action. The gap between pledges and practice was highlighted by Qu Geping, the former head of the State Environmental Protection Agency (SEPA):

> It is always my belief that what [China] lacks most [in current environmental protection and governance] is rule of law, especially strict implementation. Actually, we do not lack good guidelines and policies – we always have "leading guidelines but disjointed practice." We fail when it comes to implementation.[3]

In this chapter, I will briefly review the implementation of environmental health measures prior to the launch of the three central action plans under President Xi Jinping. Next, I will examine the changing policy context under Xi, followed by a discussion of the role of the Ministry of

Environmental Protection (the successor to SEPA) and local governments in pollution prevention and control. I will use the 2017 campaign to hit the centrally designated pollution control targets as a case study to analyze the new dynamics of policy implementation.

POLICY IMPLEMENTATION IN THE PRE-2013 ERA

Prior to 2013, government implementation efforts in environmental health were relatively circumscribed, focused mostly on improving drinking water quality (by popularizing access to safe and clean water), improving indoor and ambient air quality (through reducing exposure to smoke from traditional cookstoves, and installing FGD systems on coal-fired power plants), eliminating iodine deficiency, and banning leaded fuels. Starting in 2005, China began attacking air pollution by investing large sums in emission control projects. Measures undertaken included: (1) the reduced use of small coal-fired boilers and the increased use of clean coal and centralized/district heating; (2) enhanced control over vehicle emissions, urban dust, and Volatile Organic Compounds (VOCs); (3) accelerated construction of desulfurization (SO_2 control), denitrification (NO_x control), and dust removal (PM control) retrofit projects in key industries. These anti-pollution policy measures are similar to those implemented by advanced industrial economies, which typically include encouraging or mandating the use of fuels that emit fewer pollutants than coal, removing the dirtiest motor vehicles and boilers, and installing pollution-control equipment on major point sources and motor vehicles.[4]

In 2006, in a move that apparently was modeled after the US Environmental Protection Administration, China's SEPA created six regional supervision centers (RSCs) in east, south, north, northwest, southwest, and northeast China. These RSCs reported directly to and were funded by SEPA. While they did not have the authority to issue instructions on the daily work of local Environmental Protection Bureaus (EPBs), they supervised local governments in implementing state environmental protection–related policies, rules, and standards. In 2008, while preparing for the Beijing Summer Olympics, the central government designated the MEP, which succeeded SEPA that year, as the lead agency for handling environmental health issues.[5] The MEP at that time relied on a series of specific regulatory measures to curb increased environmental pollution.[6] They included but were not limited to:

1. Total emission control (TEC): Implemented nationwide in 1996, TEC allotted aggregated emission quotas to local governments and targeted enterprises within a limited time and area.[7]
2. Environmental impact assessment (EIA): First approved in 2002, the EIA law requires companies to investigate the effect of projects on the environment before starting construction.[8]
3. Pollutant discharge permit (PDP): First implemented in 1989 as a measure against water pollution, a PDP is issued for specified discharge pollutants. EPBs were authorized to refuse to issue PDPs based on the proposed kinds and volume of pollutants to be discharged.

While these government intervention measures were important steps toward improving environmental health conditions, they tended to be piecemeal, unsystematic, and not adequately informed by research and public participation. For more than 20 years, a fundamental goal of environmental protection had been to control the total emission of pollutants. Under that objective, upper-level governments evaluated lower-level subordinates based on the latter's performance in hitting emission control targets. But reducing total emissions does not necessarily lead to a drop in the concentration level of pollutants most harmful to health. Moreover, because of the absence of effective accountability mechanisms, polluters and local regulators had strong incentives to fabricate emissions data. As a result, despite the drop in reported total emissions over time, there was little improvement in air quality, and water and soil pollution became more severe.[9]

They also opened up a new avenue for corruption. Because EIAs were carried out by government agencies (rather than third party entities), many officials took advantage of their positions by awarding permits to big polluters in return for cash or other favors. In many localities, PDPs became a mere formality, as government officials were more enthusiastic about local economic development than environmental protection.

A 2014 MEP report admitted that the PDPs "did not have teeth in implementation," but also said it would be a pity to throw them away.[10]

Moreover, policy measures were simply too general and too ambiguous, providing local officials with little guidance to follow in implementation.[11] Specific policy measures, if any, tended to generate perverse incentives. In 2005, for example, the fine for excessive SO_2 emission was only 0.63 *yuan* ($0.08) per kilogram, although it cost 4 to 6 *yuan* ($0.5–$0.75) to clean up the same amount of pollution. Finding it cheaper to pollute, industrial firms were happy to pay the fine and continue discharging the pollutants.[12] Covert and/or excessive discharge of pollutants by industrial enterprises remained common in most of China.

INSTITUTIONAL BARRIERS TO IMPLEMENTATION. Equally important, these measures did not address two intrinsic problems in China's environmental protection apparatus: insufficient authority and lack of coordination.[13] Because of their relatively low bureaucratic rank, the two lead agencies of the 2007 action plan on environmental health, the Ministry of Health (MOH) and SEPA, had little clout when negotiating with more powerful development-oriented central ministries, commissions, or state enterprises. Being the chief coordinator of enforcing pollution control measures, SEPA acted as a "principal" to ensure its local counterparts ("agents") comply with central directives and guidelines. But it did not have final say over the personnel appointments and payroll expenditures of local EPBs, which were subject to the leadership of their "horizontal" bosses (e.g., city mayors) while reporting to the immediate "vertical" superiors (e.g., provincial EPB).[14] Under this "fragmented authoritarianism,"[15] little change was possible without strong personal intervention by the central government leaders. The weak administration under Hu Jintao (2002–2012) meant that the top leaders' call for environmental protection was mostly ignored. This institutional setting not only contributed to selective implementation at the local level but also constrained the MEP's power to inspect, investigate, and advise local governments when violations of environmental laws occurred.

Given the scale of China's environmental health crisis, huge investments are needed to clean up the air, water, and soil. China doubled total spending on environmental protection between 2006 and 2010, but pollution control measures remained underfunded.[16] In the absence of a proper incentive and constraint mechanism for investment in environmental protection, rising pollution level did not lead to significant increase in local environmental spending.[17] Between 2001 and 2014, environmental protection spending as a share of GDP was 1.13 percent, much lower than the average level of 2.15 percent in 28 European countries whose pollution levels already had peaked.[18]

Under long-standing Chinese administrative practices, which were only reinforced in the post-Mao era, local government officials have had strong incentives to resist or at least drag their feet when implementing policies that might adversely affect local revenue generation or GDP growth. As numerous studies have suggested, the ability to generate economic growth or fiscal revenue was the most important determining factor in local officials' upward career mobility.[19] Research found that a city government's spending on environmental improvements was negatively associated with the probability senior local officials (local party secretary and mayor) would be promoted.[20] In addition, local government officials often were reluctant to discuss environmental health challenges publicly in their jurisdiction, as doing so would negatively affect their ability to attract external investment.

This lack of local commitment to environmental protection was compounded by the party-state dualism – the tensions at each level between the CCP (represented by party secretaries) and state institutions (represented by government heads such as mayors). Buck-passing was common. As a result, SEPA or MOH officials had to rely on politically charged campaigns to intervene in improving environmental and public health, which were not effective in reversing the rapid deterioration in environmental health.

Local government incentives for enforcing pollution control measures were complicated further by the marriage between power and wealth. Having developed a cozy relationship with local industrial firms, local government officials either turned a blind eye on pollution from these plants or acted as their "protective umbrella" when enforcing environmental protection laws or regulations. In August 2006, villagers of Hui County in Gansu province were appalled to learn that an eight-year-old boy in the county was diagnosed with lead poisoning. Suspecting a local smelting plant, villagers rushed to hospitals to have their blood lead levels tested. Nearly 1,000 of them, including 334 children, were found to have excessive levels of toxic lead in their blood.[21]

A similar incident also occurred in Liuyang, Hunan province, where cadmium from a chemical factory poisoned 509 villagers, one-sixth of the local population.[22] Local governments in both localities labeled the polluters "key protected enterprises" prior to their shutdown, and, after news of heavy metal poisoning came out, moved to cover up the incidents, trying to deny the linkage between pollution and poisoning. This led Pan Yue, Deputy Administrator of SEPA, to label the local governments as "the ultimate culprits" in the two cases.[23]

XI JINPING CENTRALIZES CONTROL

One of the biggest changes in the Xi Jinping era is the central state's rising commitment to environmental protection. Beginning in 2013, the central fiscal authorities set up a special fund for air pollution control, totaling 25.4 billion *yuan* (approximately $4 billion) during the period of the 12th Five-Year Plan (2011–2015) to support smog control in priority regions. Since 2015, environmental protection has become a top government priority. Conducting fieldwork in China, I noticed not only that government media prominently features environment-related stories on a daily basis but also that local government officials and industrial leaders make no secret of the pressures they face in complying with environment-related laws and regulations. In 2016, the central government earmarked a total of 33.8 billion *yuan* (approximately $5 billion) to support the implementation of the three action plans, including 13.1 billion *yuan* ($1.9 billion) on water pollution, 9.5 billion *yuan* ($1.4 billion) on soil pollution, and 11.2 billion *yuan* ($1.7 billion) on air pollution.[24]

In recognition of the problem of capping the total emission level, the three central action plans unveiled during 2013–2016 focus more on environmental quality control than pollutant quantity control. This new thinking led the MEP to dissolve its department of emissions control in 2016 and reorganize it into three departments that regulate water, air, and soil quality.

The government has also started to address the root causes of pollution in China: an industry structure dominated by heavy industry (particularly chemicals), the coal-dominated energy sector, and the rapid growth of automobile ownership.[25] Fundamental to the solution to environmental pollution are the restructuring and upgrading of the industrial and energy sectors, which involves switching fuels from coal to gas and relying more on electricity, shaking off excessive industrial production capacity, and significantly raising the share of new and alternative energy in the energy mix. The 13th Five-Year Plan for Energy Development sets the goal of keeping yearly coal consumption at or below 4.1 billion tons so that its share in the energy mix is kept at no more than 58 percent by the end of 2020. Meanwhile, the share of non-fossil fuel energy use is to increase to at least 20 percent by 2030. In that regard, China's solar capacity is expected to nearly triple between 2015 and 2020, with about 14 GW of solar capacity added per year.[26]

BOOSTING TRANSPARENCY AND ACCOUNTABILITY. The newly installed Minister of Environmental Protection, Chen Jining, who was educated at Tsinghua and Imperial College London with a doctorate in civil and environmental engineering, seemed to understand better than his predecessors the importance in pollution control of participation and oversight by social forces. He offered to meet with the press right after his appointment in 2015. At the press conference, he showed his appreciation for Chai Jing's exposé, *Under the Dome*, and acknowledged the importance of interaction between the MEP and the public.[27] The action plans on water and soil pollution emphasize the need to improve transparency and engage the public in policy implementation. The plan on water pollution even showed willingness to indulge the aspirations of ordinary Chinese by "providing the general public and social organizations training and consulting services on law and regulations, and inviting them to fully participate in important environmental protection law enforcement actions and investigations of major water pollution incidents."[28]

The official statements are not just empty rhetoric. There is indeed a growing emphasis on transparency and accountability in the policy process. The post-2012 action plans on air, water, and soil pollution lay out clearer and more specific targets for local officials than prior plans.

Although China continues to suppress unflattering pollution data, it has built a nationwide network of stations to monitor air quality in all 338 prefectural-level cities, with platforms created at the provincial level to disclose data online about local factory emissions. Unlike the US EPA, which relies largely on industry's self-reported data in monitoring air quality, Chinese government agencies directly monitor emissions from polluters. According to Ma Jun, Director of the Institute of Public and Environmental Affairs (IPE), China is better equipped to implement a policy of radical transparency than any other country in the world.[29] Using publicly available data, Ma's organization creates interactive maps that expose polluting firms and pressures governments to be honest about enforcement. According to Greenpeace, such a capacity to track emissions is essential to making policy and implementing data-driven regulatory frameworks.[30]

Compared to the central action plan for air pollution, the action plan for water pollution puts greater focus on the issues of accountability and interagency coordination. It includes a laundry list of the lead agencies and their responsibilities in improving water quality. According to a senior MEP official, each of the 238 measures in the action plan "is quantified, checkable, and can be used to hold officials to account."[31] The same can be said about local action plans for pollution control. Each of the 81 bullet points in Beijing's five-year plan on air quality improvement, for example, includes the name and job title of the person responsible for transforming the policy targets into reality.[32] Similarly, each of the forty-five tasks for soil pollution prevention and control in the city lists the names of the lead and supporting agencies responsible for completing the task. Hard pollution targets, aided by the democratization of air quality data, have enabled central policymakers and outsiders to closely monitor progress, which should generate additional impetus for government officials to implement environmental health-related policies.

THE RISE OF BANDWAGON POLITY. The redistribution of political power since 2012 has transformed the institutional context for environmental health policy implementation. For China watchers, the speed and scope of Xi Jinping's centralization of power has not been seen since the era of Mao Zedong. In addition to his formal positions as President,

General Party Secretary, and head of the Central Military Commission, Xi is in charge of several leading small groups (shadowy committees that often eclipse the power of more formal political structures) as well as national committees covering comprehensive reform, cybersecurity, national security, economy and finance, and national defense and military reform. The *Economist* magazine called Xi "Chairman of Everything."[33]

To be sure, Xi's power continues to be constrained by retired politicians such as former president Jiang Zemin,[34] and his pursuit of Mao-like power has met with resistance from some of his more reform-minded colleagues in the Politburo Standing Committee.[35] Still, in most arenas, he has managed to get what he wanted. And in March 2018, the National People's Congress (NPC) removed the constitutional restriction on the maximum number of terms the president can serve, which theoretically and formally allows Xi to stay in power indefinitely. In the same month, the party revealed new rules at the top, asking Politburo members to report to Xi about their work. Xi is no longer simply the first among equals at the Politburo Standing Committee; he is the paramount leader of China.

Besides abandoning the model of collective decision-making, Xi is reversing Deng Xiaoping's legacy on other fronts. Under Deng and his immediate successors (Jiang Zemin and Hu Jintao), while the principle of party leadership remained sacrosanct, the party delegated more authority to the government bureaucracy in the making and execution of economic and social policies, while becoming increasingly irrelevant in the decision-making of the private sector and social groups. Concerned that this process excessively loosened the party's grip on power, Xi moved to reclaim political dominance for the party and for himself as the party chief.[36] In March 2018, Xi reiterated the principle Mao Zedong established 56 years before that "the party, government, the military, society and schools; north, south, east, west, and center – the party leads everything."[37]

As a result, not only do party secretaries officially become the head of state institutions, universities, and government-sponsored organizations, but party branches have also been set up at private companies, foreign-funded firms, NGOs, even Buddhist monasteries.[38] In addition, the Xi era has revived the Maoist-style personality cult, which Deng Xiaoping had firmly rejected. While newly composed songs lauding Xi and pledging loyalty to him were taking to the airwaves in China, centers dedicated to the research and study of "Xi Jinping Thought" have been sprouting up in universities and research institutes across the country. China's massive propaganda apparatus described Xi as "Great Leader," "unrivaled helmsman," "the guide of the people," "A Man Who Makes Things Happen," "Servant of the Public," "Strategist Behind China's Reform," "Top Commander Reshaping the Military," "A World Leader," and "Architect of Modernization For New Era." These are terms that were almost never used to extol the virtues of Deng and his immediate successors.[39]

Tighter concentration of influential resources by Xi in a hierarchical setting has politicized the policy process further, making "getting along by going along" the assumed motive in China's officialdom. Party members are required to establish firm "political awareness, overall situation awareness, core awareness, and emulation awareness" in order to "highly and conscientiously" align with the party-center (read: President Xi) in thoughts, politics, and actions.[40] The politically ambitious have strong incentives to jump onto Xi's policy bandwagon in the hope that they will be rewarded for being early and enthusiastic supporters. As early as January 2016, then Hubei provincial party chief Li Hongzhong became one of the first high-ranking officials to support the idea that Xi be awarded the "core" status in the Communist Party, a title Xi indeed received in late October. In pledging his full support to Xi, Li said "there is absolutely no loyalty if the loyalty is not absolute."[41] Li later was promoted to Tianjin party chief, making him a politburo member.

Xi's promises of future payoffs mixed with threats of reprisal gained additional credibility in his anti-corruption campaign, now a cornerstone of his reign. The campaign has led to the downfall of hundreds of high-level leaders who were considered powerful and not easily removed, including a former politburo standing committee member (Zhou Yongkang) and four former or sitting politburo members (Bo Xilai, Guo Boxiong, Xu Caihou, and Sun Zhengcai). Sun, the youngest Politburo member of the 18th Party Congress and a leading contender for a top position in the "6th generation of Chinese leadership," was expelled from the party in 2017 for alleged transgressions against Xi and other political and criminal wrongdoing.[42] The campaign against corrupt officials is so extensive that it rivals the purge in the Mao era: during the period from 2012 to 2017 a total of thirty-five full and alternate members of the CCP Central Committee were disciplined, which is as many as in all the years between 1949 and 2012. The purge also has reached down into lower levels of government. Since Xi took power, the anti-graft campaign has led to well over one million officials being disciplined.[43] Between December 2012 and April 2018, at least 155 officials at the provincial level and above were purged, covering all 31 provinces.[44] The vast and ruthless anti-corruption drive, while enormously popular among ordinary Chinese, deeply unsettles officials at all levels, who now try to minimize political risk by looking to Xi as a bellwether. As long as Xi remains committed to the pollution control agenda, brazen disobedience or foot-dragging should be rare. Indeed, between August and September 2017, on the eve of the 19th Party Congress, all provincial-level party heads expressed their support of environmental protection while declaring their loyalty to Xi.[45]

THE RISK OF INACTION. Nevertheless, the anti-corruption campaign also has changed the rules of the game for China's officialdom in ways that raise the likelihood of policy paralysis in enforcing pollution control. Corruption in China is so rampant that it involves almost every government official,[46] which gives Xi and his henchmen the ability to lower the hammer on practically anyone who falls out of favor. Instead of setting limited anti-corruption objectives, the leadership has indicated that it will not set any endpoint for the campaign or give pardon or amnesty to corrupt officials. It also vows to pursue each corruption case "to the very bottom" (*yi cha dao di*). In the most extreme cases, those who are found guilty can lose their lives; at a minimum their political careers are over. With significantly raised stakes, officials face a highly uncertain and insecure future. This is evidenced in the rapid increase of the number of government officials who have committed suicide since November 2012, when President Xi came to power.[47] Lower-level officials, being wary about their personal security, might officially be gung-ho about Xi's preferred policy agenda. In reality, they tend to balk at making any moves that increases the likelihood of making mistakes that could affect their upward mobility. As a result, "not taking phone calls and not writing instructions on documents" has become the new normal for some government officials.[48]

In May 2014, Premier Li Keqiang, at a State Council executive meeting, assailed this phenomenon of "holding an office and enjoying all the privileges without doing a stroke of work."[49] By saying so, Li might have overestimated the attractiveness of civil service jobs in Xi Jinping's China. From the perspective of the more than eleven million government officials, Xi's anti-graft campaign also means reduced opportunities for them to access "grey income" – off-the-books gains that have become a source of official corruption over the past four decades – that now accounts for 12 percent of China's GDP.[50] Consequently, civil service jobs are losing their appeal in China, weakening the bureaucracy through the loss of many talented individuals. According to a recruitment website, more than 10,000 civil servants quit their jobs in the three weeks following the Spring Festival in February 2015.[51] Government auditors, for example, instead are joining the private sector, working on auditing risk prevention.[52]

The problem of government inaction and policy gridlock is particularly serious at the grassroots level. Because the more they do the more likely they will make mistakes, upper-level government units have strong incentives to pass the buck to the lower-level governments under the policy principle of "regulation based on location" (*shudi guanli*). In consequence, a large number of responsibilities are shifted to the grassroots level. According to a central document issued in February 2017, township governments are held responsible for providing more than forty kinds of public services, including environmental protection, public health, and food safety.[53] Yet the grassroots-level governments have neither enforcement powers nor sufficient human and financial resources to fulfill all their designated responsibilities. Inaction and "lazy governance" (*lanzheng*) thus became common at the grassroots, inevitably undermining central efforts to clean up the environment. Citing the unscrupulous emission of air pollutants by firms surrounding Beijing, one environment expert argued: "The root cause of smog in today's China is inaction of local government officials."[54]

In the absence of effective collaboration from local governments and other functional departments, environmental protection agencies often have to rely on their own resources in enforcing laws and regulations. But they do not have the required regulatory capacity to address the country's immense environmental problems. In Fujian province, for example, there was a serious shortage of environmental law enforcement officials in 2016, averaging eleven per county. Some counties only had three such officials. The whole province had only 214 environmental law enforcement vehicles, and, in some counties, there were none.[55] When I was doing fieldwork in Zhenjiang, Jiangsu province, the head of a district environment protection bureau told me that while the bureau was supposed to have sixty staff members, only eighteen people were actually at work.[56] In the meantime, the number of sources of pollution continued to increase. According to the Ministry of Ecology and Environment (the MEE is the 2018 successor of the MEP), the number of pollution sources nationwide has increased from 5.9 million in 2010 to 9 million in 2018.[57] Against this backdrop, government regulators cannot effectively prevent violations of environmental law or incidents of environmental pollution. When that happens, grassroots regulators who have nowhere to pass the buck will become scapegoats.[58] That explains why for a long time environmental protection departments were the dead end for officials' upward mobility. An environmental expert observed:

> A reality at present is that EPB heads are in an awkward situation. If they did a bad job, they would be criticized and held accountable. If they did a good job, all they got was routine praise, and in order to protect local environment, they generally would not be transferred or promoted. In addition, EPB heads face arduous tasks and dangerous responsibilities, and often sit on a volcano of risk being held responsible for accidents. A majority of leading cadres are not willing to serve as EPB heads.[59]

According to *China Environmental Daily*, of the ninety-nine provincial EPB heads who left their jobs over a nearly two-decade span prior to 2014, only one was promoted to the vice minister level and only six became city mayors or party secretaries. "A government official – no matter how excellent she is – would end her career if she is appointed to environmental protection agencies," one such official complained.[60]

That may be changing now. As environmental issues become prominent on the top leaders' agenda, the career prospects of environmental officials have improved. In the first three years after 2014, at least six environmental officials at or above the provincial level were promoted to more prominent party or government positions.[61] They included Chen Jining, who was appointed Beijing mayor in May 2017 after his three-year tenure at the MEP. His successor at the MEP, Li Ganjie, was vice minister of environmental protection before being appointed Deputy Party Secretary of Hebei province half a year before. Their promotions, according to a Chinese scholar, suggest that "in the future, officials of local environmental protection agencies will no longer be marginalized but will have a good chance to be transferred to other important posts as leaders."[62]

THE ENVIRONMENTAL COORDINATION MESS

In China's "regime complexes" (i.e., sets of overlapping institutions governing a common issue area),[63] one of the main challenges in policy implementation is to develop effective mechanisms to coordinate and restructure the often-conflicting interests of a multitude of bureaucratic actors with conflicting agendas. The "negotiation constraints"[64] are particularly an issue for the MEP, whose administrative rank is no higher than other central ministries or commissions. In implementing a clean energy policy, for example, it is important to have cooperation from National Development and Reform Commission (NDRC), Ministry of Housing and Urban-Rural Development (MOHURD), and National Energy Administration (NEA) over issues including energy price adjustment and infrastructure improvement.

But that's only the beginning. Effective action also hinges upon the support of local governments and huge state-owned enterprises such as Sinopec, about whose business the State-owned Assets Supervision and Administration Commission (SASAC) also would have a say. Regulations to audit outgoing officials' compliance with regulations rely on active engagement by the Ministry of Finance (MOF), which has little experience working on environmental issues.[65] Furthermore, the MEP has to make sure subnational governments that belong to the same "pollution clusters" (e.g., Beijing-Tianjin-Hebei) coordinate with each other in hitting emission reduction targets. The effectiveness of Beijing's campaign against air pollution, for example, depends on how effectively its surrounding regions fight the same problem. According to a study released by Beijing's EPB, the municipality's surrounding regions contributed over 50 percent of the pollutants on heavily polluted days when the AQI exceeded 200 in Beijing.[66]

By 2015, it was clear that many of the objectives identified in the 2007 action plan on environmental health, such as setting up an environmental health leading small group and constructing an information sharing system for environment and health departments, did not materialize. The MEP did manage, however, to build a nationwide surveillance network for air quality and put in place a policy framework for addressing environmental health-related issues. Equally important, since 2013 the MEP (now the MEE) has played a central role in drafting and implementing the three action plans for pollution control.

In order to achieve the post-2012 action plan goals, the central government, through the MEP, divided large goals into numerous sub-goals, and each sub-goal would involve two to eight government ministries. Table 4.1 shows the designated lead agencies and participating agencies for each of the goals on water pollution control. The MEP is the designated lead agency (*qiantou*) for implementing all ten goals. All but one of the goals, however, involves more than two lead agencies and over six participating agencies. In total, seventy-two of the seventy-six sub-goals require interagency coordination at the national level.[67]

The same is true for implementing the action plan on soil pollution. Of the fifty-one policy areas listed in the action plan, MEP is the sole lead agency for only twenty of them. Seven other central ministries or commissions (MOA, MOF, NDRC, MOST, MWR, MLR, and MOHURD) are the sole lead agencies for another fifteen sub-goals (Figure 4.1). Eleven of them require the MEP to work with other central agencies.[68]

The MEP's prestige and power in China's crusade against pollution rose in the post-2012 era, but it suffered several blows because of the smog crisis in 2013 and the lack of progress in significantly bringing down the PM2.5 level under Zhou Shengxian, who started in his job in 2005. In

TABLE 4.1 *Water pollution prevention and control goals, and lead and participating central agencies*

Goals	Lead central agencies	Participating central agencies
Comprehensive emission control (industrial pollution control, sewage disposal, rural pollution control, port pollution control)	MEP, MOHURD, Ministry of Agriculture (MOA), Ministry of Water Resources (MWR), Ministry of Transport (MOT)	Ministry of Industry and Information Technology (MIIT), MOHURD, MOA, MEP, MWR, Ministry of Land and Resources (MLR), NEA, Ministry of Science and Technology (MOST), Ministry of Commerce (MOFCOM), General Administration of Quality Supervision, Inspection and Quarantine (AQSIQ) (10)
Economic structural change (industrial structural adjustment, environment access, space layout and planning, circular development, recycled water use)	MIIT, MEP, NDRC, MOHURD	MLR, NDRC, MEP, MOHURD, MWR, NEA, MIIT, MOT, State Oceanic Administration (SOA) (9)
Science and technological support (development and promotion of related technologies, development of environmental protection industry)	MOST, NDRC, MOF	NDRC, MIIT, MEP, MOHURD, MWR, MOA, SOA, MLR, MOST, MOF, National Health and Family Planning Commission (NHFPC) (11)
Making full use of market mechanisms (optimizing taxes and fees, financing from multiple sources, introducing financial incentives)	NDRC, MOF, State Administration of Taxation (SAT), People's Bank of China (PBOC)	MEP, MOF, MOHURD, MWR, NDRC, MIIT, MOFCOM, AQSIQ, General Administration of Customs (GAC), the China Insurance Regulatory Commission (CIRC), China's Securities Regulatory Commission (CSRC), China's Bank Regulatory Commission (CBRC) (12)
Environmental enforcement, surveillance, regulation, and risk management	State Council Legislative Affairs Office (LAO), MEP	NDRC, MIIT, MLR, MEP, MOHURD, MOT, MWR, MOA, NHFPC, CIRC, SOA, AQSIQ, Ministry of Public Security (MPS), State Commission Office for Public Sector Reform (SCOPSR), State Administration of Work Safety (AWS) (15)
Water ecological safety (drinking water safety, aquaculture development, water treatment)	MEP, SOA, MOHURD, MOA	NDRC, MOA, MOHURD, MWR, NHFPC, MOF, MLR, MOFCOM, MOT, MEP, MOA, MIIT, Ministry of Foreign Affairs, State Forestry Administration (SFA) (14)

TABLE 4.1 *(continued)*

Goals	Lead central agencies	Participating central agencies
Responsibility fulfillment and performance appraisal, including use of appraisal results for fund allocation	MEP, MOF, NDRC	NDRC, MOF, MOHURD, MWR, MOST, MIIT, MOA, SOA, SASAC, MEP, Central Organization Department, Ministry of Supervision (12)
Public participation and societal oversight	MEP	NDRC, MOHURD, MWR, NHFPC, SOA, MIIT (6)

Source: Author's Database.
Note: Leading and participating central agencies are for sub-goals only. Since one policy goal may include several sub-goals, a central agency can be both lead and participating agencies under a policy goal.

late November of 2014, after the rare blue sky in Beijing during the APEC summit gave way to lingering smog in the city, Zhou was summoned by central inspectors for a meeting. Two months later, he was replaced by Chen Jining, then president of the prestigious Tsinghua University.

Chen climbed the professional ladder at his alma mater in China before being appointed the Minister of Environmental Protection.

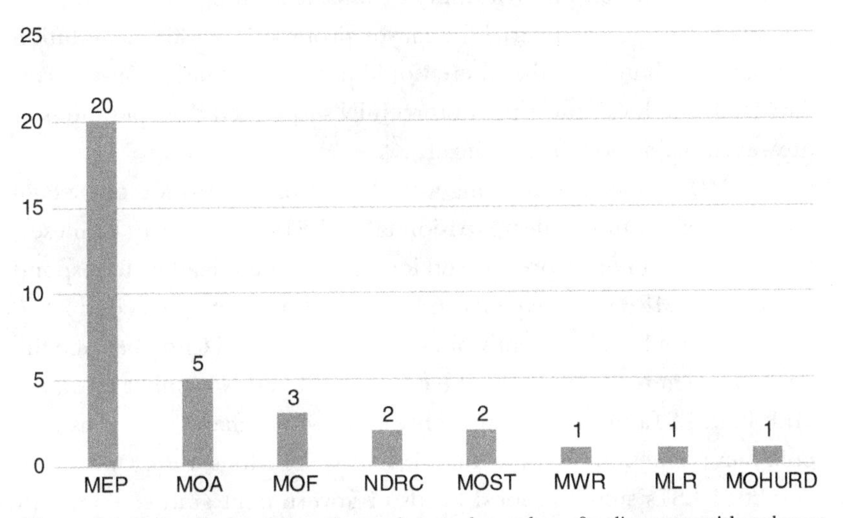

Figure 4.1 Soil pollution prevention and control: number of policy areas with only one central lead agency.
Source: Ministry of Ecology and Environment, June 1, 2016. www.mee.gov.cn/home/ztbd/rdzl/trfz/ss/201606/t20160601_353133.shtml
Note: MEP=Ministry of Environmental Protection; MOA=Ministry of Agriculture; MOF=Ministry of Finance; NDRC=National Development and Reform Commission; MOST=Ministry of Science and Technology; MWR=Ministry of Water Resources; MLR=Ministry of Land and Resources; MOHURD= Ministry of Housing and Urban-Rural Development

While Chen's impressive academic credentials were important for his promotion, his connections with powerful members of the ruling elite perhaps mattered more. At Tsinghua, he had significant overlaps with Chen Xi, who happened to be a close ally of President Xi. It is known, for instance, that Chen Xi was Xi Jinping's roommate at Tsinghua in the late 1970s. Chen Jining was promoted to the minister position when Chen Xi was Executive Deputy Minister of the Central Organization Department in charge of the appointment of other minister-level officials. To put it simply, Chen Jining is connected to President Xi through their mutual association with Chen Xi.[69]

Minister Chen lost no time demonstrating his support for President Xi's anti-pollution agenda. In a move that showed his dissatisfaction with the MEP's role in policy implementation, Chen vowed to "flip over the 'soft' enforcement of environmental protection law, and make law compliance the new normal."[70] Just three days before Chen made the remarks, MEP inspectors summoned the mayor of Linyi, one of the most polluted cities in Shandong province, urging him to crack down on firms in violation of environmental laws. Admitting that he came to the talk with a "heavy heart," the mayor promised to address pollution with "an iron hand," so that there would not be a second such talk. Five days later, the local government forcefully suspended the operations of fifty-seven major polluters in the city.[71]

The MEP inspectors' meeting with the mayor of Linyi was one example of the environmental supervision talks (ESTs or *yuetan,* in Chinese), which seek to generate pressure on local government leaders to respond to and meet MEP requirements and deadlines. First introduced in 2007, ESTs were not held frequently or taken seriously until Chen became the head of the ministry. Between the end of 2014 and November 2015, the MEP held ESTs in more than twenty cities where major concerns over policy implementation were raised. Unlike previous *yuetan* practices, post-2014 ESTs summon local heads of government (rather than only their associates) and encourage media coverage. The MEP now carries big sticks when meeting with local officials. It may halt the EIA approval process for new pollution emission projects, as it did in Linfen, Shanxi province, in January 2017. The practice also can lead to the reprimand and punishment of local officials. For example, in localities covered by the MEP's northern China regional supervision center fifty-seven local officials were reprimanded, warned, or removed following ESTs in the first seven months of 2015.

CATCH-22 FOR LOCAL GOVERNMENTS

The primary targets of ESTs are local governments, which all three central action plans identify as the main body of policy implementation. But the central government, by expecting local officials to continue promoting economic growth (which has historically resulted in more polluting activities) *and* hit newer environmental protection targets without appropriate financial, political, and moral support, has placed local governments in a Catch-22 situation.

While the central government is committed to environmental protection, many local government leaders still view economic growth as the key to achieving their career advancement. "It's very difficult to correct once and for all the urge of local government officials to pursue work performance in this area," one Chinese scholar told me.[72]

The difficulty in changing the local government mindset is conceded by central government–directed media: "In some localities, government officials no longer 'judge a hero by GDP' in rhetoric, but they still do so in their thoughts, their practice and their hearts."[73] Government media reported that the Liaoning provincial party secretary did not convene any meetings on environmental protection during 2013–2014 while Anhui provincial government reduced the weight of environmental indicators in government officials' performance evaluation.[74] Indeed, twenty of the thirty-one provincial and municipal governments gave the lowest priority to environmental protection in their work reports issued in early 2017, and Beijing was the only one that ranked pressures of environmental protection as the no. 1 issue in the "problems and difficulties" section.[75]

Not all local governments are reluctant to invest in environmental protection, however. Research found that fiscally weak localities tend to be slower to implement centrally mandated pollution control measures for fear of undermining their existing base of revenue. In fact, they have strong incentives to attract polluting industries that, in rural areas, correlate highly with cancer villages. Fiscally strong localities, by contrast, have more incentives to comply with central environmental protection laws and regulations because officials there face less pressure to make revenue generation their sole focus, even though they also aspire to increase revenues and advance their careers.[76] The fiscal disparities might explain why among all China's provincial units, only Shanghai and Guangdong – two provincial units with high fiscal capacities – explicitly proposed to increase the share of environmental protection spending to 3 percent of their GDP from 2011 to 2015.[77] It might also explain why cash-strapped local governments in Inner Mongolia (an autonomous region of northern China) did not take any major action on the serious pollution caused by industrial firms at Tengger Industrial Park in 2014 until after Xi Jinping three times made written comments on the issue and the State Council sent special inspection teams to urge a change.[78]

To try to ensure that local governments follow central directives loyally and competently, the State Council asks provincial governments to sign a "target responsibility pledge" (*mubiao zeren shu*), which they then pass down to lower-level government officials in charge of pollution control. In this way, targets are assigned to bureaucratic units at different levels (e.g., province, prefecture-level city, district or county or county-level city). Table 4.2 shows how the Beijing municipal government set up and broke down its PM2.5-related pollution control targets. The government divided the capital's seventeen districts into four groups based on their potential to realize improvement in air quality. Tier 1 districts (Huairou, Miyun, and Yanqing) were rural counties where "ecological conditions were relatively good."[79] They were ordered to reduce the PM2.5 concentration to 50 $\mu g/m^3$. By contrast, Tier 4 districts (Mentougou, Fangshan, Tongzhou, Daxing, and the Beijing economic development zone), where industrial firms were concentrated, were asked to drop PM2.5 level to 65 $\mu g/m^3$. Similar arrangements were made along practical lines, with tasks prioritized into coal use reduction, vehicle emission control, and industrial pollution control.[80] The implementation will be subject to appraisal, and the results will be the basis for evaluating officials' performance and a reference for distributing future environmental protection funds.

TABLE 4.2 *Beijing air quality improvement: target management*

Overarching target	25 percent PM2.5 concentration drop over 2012 level; PM2.5 reaching 60 $\mu g/m^3$ annual average by 2017			
Tier	1	2	3	4
Targeted % drop in PM2.5	>25%	>25%	>30%	>30%
Annual average ($\mu g/m^3$)	50	55	60	65
Districts	-Huairou -Miyun -Yanqing	-Shunyi -Changping -Pingguo	-Dongcheng -Xicheng -Chaoyang -Haidian -Fengtai -Shijingshan	-Mentougou -Fangshan -Tongzhou -Daxing -Beijing Economic and Technological Development Area
Target reassignment among functional departments	1. coal consumption reduction: Municipal Management Committee; Municipal Committee of Economy and Informatization; Municipal Agriculture Committee; EPB; Administration for Industry and Commerce 2. vehicle emission control: EPB; Municipal Committee of Commerce; Municipal Commission of Transportation; Bureau of Public Security; Municipal Commission of Housing and Urban-Rural Development; Municipal Bureau of Water Affairs; Municipal Bureau of Landscape and Forestry; Municipal Committee of Economy and Informatization; Beijing Science and Technology Committee; Urban Planning and Land Resources Committee 3. industrial pollution control: EPB, Municipal Commission of Housing and Urban-Rural Development; Municipal Management Committee, Municipal Committee of Commerce; Committee on Economy and Informatization; Food and Drug Supervision Bureau.			

Source: "Beijing shi 2013–2017 nian qingjie kongqi xingdong jihua zhongdian renwu fenjie" (Beijing Municipality Air Cleaning Action Plan (2013–2017): Priority Task Decomposition), www.d1ev.com/news/zhengce/21452; Huang Han, "Zhibiao zhili jiqi kunjing" (Performance Target Management and Its Dilemma), *Haerbing gongye daxue xuebao* (Journal of Harbin Institute of Technology), Social Sciences Edition, 18, no. 6 (December 2016), 37–45.

In essence, the performance evaluation in environmental protection is no different from the cadre evaluation system introduced in the post-Mao era, which assesses local officials' work based on certain key performance indicators. In fact, since 2013, PM2.5 control has been included in the cadre performance evaluation system and subject to the so-called one vote down (*yi piao fou jue*): if the pollution control target is not hit, the local government leadership in question will be denied an "advanced" assessment and the party secretary or government head can be blocked from promotion, no matter how well other work is completed.

That sets up another Catch-22 situation, rendering effective policy enforcement difficult, if not impossible. First, absent public participation in the policy process, the targets often are set in such an arbitrary way that they are neither popular nor feasible. Second, when facing myriad, sometimes conflicting targets, government officials may choose to work on only certain targets, especially those that are quantifiable and clearly reflect top leaders' priority, to the detriment of other ones. Third, local officials striving to hit the targets and make them look good in performance review may turn to unusual, even unlawful means – a blanket ban, overkill, and fabricating data – which generate outcomes that defeat the purpose of the system.

An interesting policy development in China's pollution control is the phenomenon of *cengceng jiama*, or imposing additional targets and requirements at every lower administrative level. As Table 4.2 shows, even though Beijing municipality was asked by the central government to curb PM2.5 concentration by 25 percent in a five-year period and reach the annual concentration average of 60 $\mu g/m^3$, each district was hit with even stricter requirements: Tier 1 and Tier 2 districts were to reach PM2.5 concentration annual average of 55 $\mu g/m^3$ or lower, while Tier 3 and Tier 4 districts were to curb the PM2.5 level by more than 30 percent. *Cengceng jiama* is seen, too, in the treatment of the so-called black-odor rivers (*heichou shuiti*). While the central government asked local officials to complete their task in three to five years, the authorized implementation period shrank to two to three years at the provincial level. When it reached the grassroots level, only one year was left to get the job done. This raises a feasibility issue in policy implementation. As one expert pointed out, it would take about sixty years to control thirty years of pollution in other countries, but in China local governments were asked to complete the job within only one to two years.[81] The top-down amplification of pollution control targets along the jurisdiction levels mirrors the pattern of setting economic growth targets, suggesting the rampant use of *cengceng jiama* in policy implementation in China.[82]

By official accounts, *cengceng jiama* should be viewed as calculated institutional response to the problem of *yali chuandao cengceng dijian*, or pressures of implementation being decreased progressively level by level in transmission. As an article in the MEP's official newspaper contended, the central leadership's decisions on environmental protection need to be carried out level by level and the pressures need to be transmitted level by level, yet,

> When some local party committees/governments and related agencies are involved in enforcing environmental protection policies, there exists the problem of illegally "bending rules" and "watering down [policies]." [Local officials] either lower environmental protection standards, even do not comply with certain national industrial regulations, or cut those assigned to implement their own initiatives some slack. These foot-draggers are reluctant to or dare not assume responsibilities. They bypass difficult issues and carry out central environmental protection policies selectively.[83]

The reality, as a government newspaper observed, is that the lower the bureaucratic level, the more likely policy enforcement is to encounter resistance from local and departmental interests. Given that policy success hinges upon implementation at the grassroots level, the attenuation of pressures from above could lead to implementation failure unless stronger countermeasures are undertaken.[84]

Fundamentally, *cengceng jiama* is a byproduct of the authoritarian, hierarchical setting that governs policymaking and implementation in China. First, upper-level governments have strong incentives to impose ambitious, often impossible, targets on lower levels. Second, since Chinese government officials at each level of the administration all are appointed by higher authorities, they are accountable to their immediate superiors rather than to the people. Due to this upward accountability, government officials may act on their own to raise the bar even further as a way to fawn their bosses and demonstrate their loyalty. Third, under the "promotion tournament model,"[85] the policy appraisal and cadre assessment system has been designed in a way that rewards those who have done a better job in displaying achievements and hitting quantified targets. This competitive dynamic encourages government officials at all levels to design ambitious, sometimes unrealistic policy targets, which only expands the discrepancy between policy aspirations and implementation.

Still, government pollution-control endeavors have resulted in occasional institutional innovations. Under Xi, China popularized the practice of appointing local heads of government as "river chiefs" (*hezhang*) to improve transboundary river governance, including coordination and accountability in preventing pollution and improving water quality.[86] The practice, first initiated in Wuxi city, Jiangsu province, after a blue algae outbreak in Taihu Lake in 2007, requires top officials from the provincial, city, county, and township levels to be responsible for the rivers under their jurisdiction. A river chief finding signs of pollution on the section of the river for which he is held responsible is obliged to coordinate with various functional departments (e.g., environment, water resources) to solve the problem. Officials who arrive at annual goals are rewarded, and those who are found to have neglected their duties are penalized with fines and loss of promotion opportunities. This system aims at fixing the problem of buck passing in policy implementation, i.e., government officials shirk their responsibilities over pollution control and shift the blame onto other jurisdictions.

Since 2015, more institutional mechanisms have been set up to delineate the responsibilities and improve accountability. In July 2017, the Central Leading Group for Deepening Overall Reform approved a trial of the "Pilot Program on Conducting Natural Resource Assets Departure Audit on Leading Cadres" and "Measures for the Accountability of Party and Government Leaders for Damage to the Ecological Environment." The two documents established "responsibility for life" rules, under which those government officials whose decisions lead to serious environmental degradation are held accountable even after they leave office. The central leading group also approved a trial of "Measures on environmental protection inspection," which formally introduced "party-administration shared responsibility" and the "one-post-two-responsibilities" systems. Under the "party-administration shared responsibility" system, local party leaders and government heads share responsibilities for pollution control in their jurisdiction. Under the "one-post-two-responsibilities" system, heads of functional departments other than EPBs are required to assume responsibilities related to pollution control in addition to their other regular work. For example, a local propaganda head would be held responsible for the assigned environmental protection work *and* regular operational work in her portfolio (e.g., being in charge of press and publication).

In addition to these horizontal institutional changes, the central government also has sought to enhance the authority of the MEP vis-à-vis local governments. In 2017, MEP's regional supervision centers were converted from "service units" (therefore not considered part of the state nomenklatura or "*bianzhi*") to one of MEP's formal dispatched agencies called environmental supervision bureaus, thereby being fully integrated into the MEP. The regional environmental supervision bureaus and provincial EPBs are of the same bureaucratic rank, but the former are more powerful in that they enforce environmental laws and regulations on behalf of the MEP. Since 2016, in order to reduce local governments' interference in the enforcement of environmental laws and regulations, the central government also has pursued more centralized vertical management of environmental monitoring and oversight below the provincial level. Under the vertical management structure, the EPB in a city will still be subject to "dual leadership" by the provincial EPB and the city government, but provincial EPBs now will have primary authority over the appointment of the leading staff members of the local EPB. As part of the institutional change, the initial supervisory functions assumed by city and county EPBs now are returned to the provincial EPB, which dispatches supervisory organs to the sub-provincial levels and sets up an ombudsman system to strengthen the provincial supervision over local government officials. With the blessings of President Xi and Premier Li, this reform has been promoted in all provinces since November 2018 and was to be completed by March 2019. The reform was implemented in tandem with the conversion of county-level EPBs to agencies dispatched by city EPBs, with the latter responsible for the former's funding and staffing.

Furthermore, in order to reduce potential local interference in emissions data, the MEP began in November 2016 to bring under its direct central control all 1,436 air quality monitoring stations set up across the nation. There is now a new division of monitoring-related responsibilities: city EPBs conduct assessment work while county EPBs are only responsible for law enforcement-related monitoring. The ultimate objective is to centralize control over air, water, and soil quality monitoring.[87]

The enhanced authority of the environmental protection bureaucracy, beneficial as that may be, raises concerns about policy implementation at the local level. MEP regional centers, it is argued, sometimes interfere with the local governments exercising their legitimate authority.[88] One local environmental official noted that after local EPBs were placed under the primary leadership of provincial EPB, local governments would lose their institutional "handle" in implementing environmental protection measures, while local EPBs also would encounter difficulties in policy implementation in the absence of adequate support from local governments.[89] After all, the environmental law holds local governments ultimately accountable for environmental quality under their jurisdiction. One environmental researcher asked: "If they are deprived of key enforcement authority, how can they be accountable for what they are unable to control?"[90] Mayors after being summoned for the environmental supervision talks may vent frustration on local environmental protection officials who, in turn, become scapegoats. Indeed, of the seven city government officials disciplined after the EST in Zhumadian, in southern Henan province, four were environmental protection officials.[91]

THE 2017 CAMPAIGN

The year 2017 was the "ending battle" for hitting the targets set by the action plan for air pollution. In order to hit the year's targets, China needed to cut 2012 levels of PM2.5 by more than a quarter in the Beijing–Tianjin–Hebei (BTH) region and bring average concentrations in the capital down to around 60 $\mu g/m^3$. That was no easy job, especially after the near record-high smog in January and February caused the rebound of PM2.5 levels. In the first half of 2017, BTH PM2.5 averages rose 14.3 percent over the same period in 2016.[92] Some Chinese environmental scholars expressed concerns over China's ability to meet the stipulated policy targets. One predicted that it would take another two to three years for Beijing to hit the assigned targets on the grounds that the city had exhausted almost all of its routine measures, and additional measures would not have immediate impacts.[93]

Nevertheless, the central government was determined to beat the odds. As the deadline drew near, it geared up its pollution control efforts. In January, the National Energy Administration (NEA) announced that it would cancel 103 coal-fired power plants that were planned or under construction, which would eliminate 120 gigawatts of future coal-fired capacity.[94] The same month, the Beijing municipal government announced its plans to establish a special police force to crack down on persistent environmental offenses. By the end of the year, the environmental police force had been set up in nine provinces.[95]

In March, the central government released its "2+26" plan for the winter. Covering the period from mid-November 2017 to mid-March 2018, the new plan was set up hastily to further bring down pollution in Beijing, Tianjin, and 26 outlying cities in the smog-plagued provinces of Hebei, Shandong, Henan, and Shanxi. In response, the Beijing municipal government announced that it would eliminate small coal-fired boilers, in an attempt to achieve zero coal use in six major districts, and cut coal consumption by a further 30 percent in 2017.[96] The municipal government spending on air pollution control reached 18.2 billion *yuan* in that year, a 10 percent increase over 2016.[97] In August, the MEP asked that average PM2.5 concentrations be cut by more than 15 percent between October 2017 and March 2018 in the BTH region. Under the 143-page winter smog "battle plan," local air quality monitoring stations not managed by the MEP would be placed directly under the control of provincial EPBs, which would rank counties and cities based on their air quality level. Those industrial firms identified as "scattered, chaotic and dirty" (*san luan wu*) would be closed down by September. Furthermore, the 2+26 cities were to switch at least three million homes from coal to gas or electric heating by the end of October, three times more than the number completed in 2016.[98] The government moved to set up no-coal zones in cities around Beijing, where factories as well as households would no longer be allowed to burn coal. The new plan required Beijing, Tianjin, and two cities in Hebei province (Langfang, approximately midway between Beijing and Tianjin, and Baoding, approximately 93 miles southwest of Beijing) to complete the setup of the no-coal zone by the end of October. That meant that 44,000 small coal-fired boilers would have to be shut down in less than one year, compared to the 10,000 such boilers shut down in 2016.

Meanwhile, the rewards and punishments for local government officials were increasingly linked to results. In February 2016, the State Council set up a leading group on environmental protection supervision work. Headed by executive vice-premier Zhang Gaoli, the group is in charge of directing and coordinating environmental protection supervision. Five months later, the MEP launched the practice of central environmental protection inspection (CEPI) to supervise local pollution control work. According to a senior government official, "General Secretary Xi promoted the introduction of CEPI in person: he made important written comments and instructions for every key link and in every pivotal moment, and reviewed each supervision report."[99]

In contrast to previous environmental supervision tours, which focused on investigating polluting firms, CEPI emphasizes making officials answerable for their actions and ensuring their cooperation in implementing environmental laws and policies. With support from the CCP disciplinary and organizational agencies, the MEP demonstrated that it has teeth when it comes to pollution control. Following an investigation in the summer of 2016, it was revealed in November 2017 that central inspectors had disciplined 1,140 officials in eight provinces, including 130 department-level officials (*ting*) and 504 division-level officials (*chu*).[100] According to a leading environmental scholar in China, the impact of such disciplinary actions should not be underestimated: "What government officials are most afraid of is that their career prospect is negatively affected [by disciplinary actions]; admonishing, disciplining party members, even public apology can have a major impact on an official's political career."[101]

In September, Wang Sanyun, former Party Chief of Gansu province, was expelled from the party for taking bribes and being "passive" in implementing the party's key policies, including ignoring Xi's repeated instructions to stop the ecological damage to the Qilian Mountains.[102] The unprecedented number of officials being disciplined sent a strong signal that failure to take environmental policy implementation seriously would cost government officials their careers. In April 2017, the ministry also began the largest inspection campaign on record when it deployed some 5,600 environmental inspectors to the BTH region and nearby areas to check the compliance of environmental laws and policies in the 28 northern cities.[103] According to Minister Li Ganjie, the MEP would rank the cities based on how their air quality had improved by March 2018. For those in the bottom three ranks, the following government officials would be held responsible: vice mayor (if the improvement was less than 60 percent of the targeted level), mayor (if the improvement was less than 30 percent of the targeted level), and party secretary (if the air quality was still getting worse).[104]

The number of inspectors increased to 7,000 in September, "the time of toughest air pollution regulation in history."[105] By the end of 2017, central environmental inspectors were dispatched to all 31 provinces, municipalities, and autonomous regions. Since the practice was piloted in Hebei province in 2015, 29,000 firms have been placed on file for punishment, 1,518 cases have been placed on file for investigation, 1,527 people have been detained, 18,448 local government leaders have been subject to EST, and more than 18,000 officials have been disciplined.[106] In a nutshell, the campaign was pursued on such a large scale with vigor because its architects hoped to transmit the pressures from above to the lowest levels of government without obstruction. A full evaluation of the 2017 campaign will be conducted in the next chapter, but there is no doubt the campaign has enabled the state to improve its control over pollution by a significant measure.

Industrial firms that were previously the darlings of local states now had to take the brunt of the environmental protection storm. In the morning of a summer day of 2017, I was invited to have breakfast with a friend who was well respected in the locality. He was joined by Mr. G, a shrewd local businessman who happened to own the restaurant. We sat at a private room reserved for Mr. G. After some chit-chat, Mr. G turned to my friend, complaining that he had been losing money every day after local environmental protection bureau suspended the operation of his alloy steel factory. "I don't understand why they did that. Could you do me a favor and ask the EPB people to bend the rules?" he asked. My friend refused: "The heat is on now. There is not much I can do."

NEW DEVELOPMENTS

Even as the air pollution action plan expired at the end of 2017, China stepped up the fight against pollution. Beijing's municipal government proposed new goals to reduce PM2.5 concentration to 56 $\mu g/m^3$ in 2020 and achieve "fundamental improvement" in air quality by 2035.[107] In January 2018, a uniform environmental protection levy came into effect. Aimed at changing industrial firms' behavior through a market-based incentive, the new system charges polluters nationwide 1.2 to 12 *yuan* ($0.18-$1.8) for every 0.95 kg of NO_x and SO_2 they discharge.[108]

In March, the Ministry of Environmental Protection was superseded by the Ministry of Ecology and Environment (MEE). While inheriting most of the responsibilities of MEP, MEE saw its remit expanded to incorporate pollution-related functions from NDRC, MWR, and SOA. As the country's top emission regulator, MEE is in a more powerful position than MEP to address issues of inter-ministerial coordination and central-local implementation gap.[109] In July, the government unveiled the Three-year (2018–2020) Action Plan for Winning the Blue Sky War. Compared to the earlier air pollution action plan, the new plan applies to all cities that are prefectural or higher level, and mandates at least an 18 percent cut in PM2.5 levels on a 2015 baseline.

The expansion puts more competitive pressure on subnational governments. In view of the better performance of Zhejiang and Beijing in reducing PM2.5 concentration, Jiangsu province launched four rounds of "special actions" as part of its Blue Sky Protection Campaign in 2019.[110] Local governments also began to be more serious about water pollution control. In my hometown, after national media exposed the lack of progress in improving the city's water quality, the local government unveiled a plan in October 2018 on accelerating treatment of water pollution. Villagers were encouraged to report their concerns to the local government. "It took just a phone call for them to send down people to treat the black-odor river in front of my house," my nephew was thrilled to tell me when I visited in spring 2019.

Beyond the drive against air, water, and soil pollution, the government has also kicked off campaigns to confront previously ignored environmental health challenges, such as garbage control. One of the objectives is to raise China's trash recycling rate from under 20 percent to 35 percent in 46 Chinese cities by 2020. The policy process was reminiscent of the national response toward Chairman Mao's comments five decades before that "barefoot doctors are good."[111] During his inspection tour in Shanghai in November 2018, Xi Jinping said, "garbage sorting work is the new fashion!" Shanghai municipal leaders took the cue from Xi by announcing two months later that the city would roll out a mandatory garbage sorting system by July 2019.

Considered "the most comprehensive waste management regulations to be applied on such a scale anywhere in China," the new regulations requiring sorting into recyclable waste, hazardous waste, residual waste, and kitchen waste immediately raised critical concerns on how compliance will be monitored and enforced.[112] Both sticks and carrots are used to induce compliance: while individuals who do not correctly separate their trash are to be fined 50 to 200 *yuan* ($7 to $29), those who have excelled in the process would be rewarded with redeemable points. The city government also distributed flyers to every family, displayed posters around the city, and deployed volunteers to supervise the trash sorting.

Shanghai is setting the model for the rest of the country. In February 2019, the Ministry of Housing and Urban-Rural Development (MOHURD) required that all cities at the prefectural level or above launch household trash sorting in the year and put in place a nationwide urban garbage sorting system by the end of 2025. Beijing, Guangzhou, Shenzhen, and Hangzhou followed in Shanghai's footsteps, enacting or revising regulations on garbage classification and adding more penalties for violations.[113]

FRAGMENTATION AND INTEGRATION IN POLLUTION CONTROL

When Xi took the helm in 2012, one of his major concerns was how to connect the wheel to the rudder: eliciting compliance from lower levels. Driven by an increasingly individualized society and growing coordination and monitoring problems in the officialdom, centrifugal forces resistant to central control were overwhelming strong. There was a saying that *zhengling buchu Zhongnanhai* ("Government decrees are not honored outside of the Zhongnanhai compound"), which spoke to the ineffective central leadership. Unlike previous party leaders (Deng, Jiang, Hu), Xi rapidly recentralized his political power to a level that rivals Mao's. The recreation of the Maoist bandwagon polity reshaped the institutional contours of policy implementation in the Xi Jinping era.

Pollution control was a key element of this shift. The central leaders pushed for the enactment of three pollution-related action plans and sought to make policy measures less general and more operational. With the elevation and embedding of environmental health issues, the power and prestige of central environmental protection agencies have been strengthened. This has been facilitated by the introduction of new policy instruments and mechanisms (e.g., CEPI) aimed at improving accountability and policy coordination. Such policy measures, when combined with other integrative mechanisms – leadership commitment, party discipline, and personal ties – were aimed at mitigating implementation problems and allowing policies proclaimed in Beijing to bear resemblance to the reality at lower levels.

As this chapter shows, however, many of the inherent policy implementation problems continue to be left untouched. They include upward accountability, lack of public participation, conflict between functional departments and territorial governments, and the central state's inability to effectively monitor and evaluate bureaucratic performance. The conflict between development goals and environment protection and the anticorruption campaign exacerbate policy inaction and foster selective implementation. In the face of sustained forces of centrifugation and fragmentation, how successful has the central government been in pollution control? This will be the subject of the next chapter.

CHAPTER 5

An Assessment of Policy Effectiveness

As the Boeing 777 took off from Newark for Beijing in mid-December of 2017, I realized that I had forgotten to pack my N95 respiratory mask. I regretted it immediately. Upon approaching the landing some fourteen hours later, though, my spirits lifted. As we descended, I could see blue sky over China's capital and, when I got off the plane, it was indeed a clear, if chilly, day. When asked whether this was just an aberration from the norm, my taxi driver told me proudly that the city was experiencing fewer smog days than just a couple of years before.

Awesome, I thought – until I came to understand the huge, unintended price people living in the vicinity of Beijing were paying for the improvement in the capital's air quality. Striving for blue skies, local authorities in northern China were pressed, in 2017, into converting coal-burning heating and cooking to cleaner fuel-burning systems. The implementation of the policy went *too well*, however, creating shortages in gas in some regions that left millions without proper heating in freezing temperatures. Alerted by the news that elementary school students without indoor heating were forced to take classes outdoors in the sparse sunlight to warm themselves, the Ministry of Environmental Protection (MEP) took a policy U-turn, allowing some northern cities to continue burning coal for heat. Unfortunately, by the time the U-turn was approved, a large number of coal burners had already been dismantled.

This is but one example of the mixed outcomes of China's evolving environmental policy, outcomes that raise questions about the success of pollution control efforts under President Xi Jinping. In this chapter, first, I will use the typical method of policy assessment to examine the congruence between the policy goals enshrined in the three central action plans and the actual outcomes. Next, I will discuss whether declarative policy was implemented efficiently and in ways that conform to the procedures called for in the policy documents. My analysis will allow us to take into account the intended – and the unintended – results from China's newly aggressive environmental policy.

THE SUCCESS STORY

China has made impressive progress in tackling environmental health, especially in controlling air pollution. Five years into the 2013 central action plan for tackling air pollution China managed, by government accounts, to achieve nearly all its goals, which included an overall improvement in air quality and a reduction in the days of worst pollution "by a large margin" nationally, as well as "marked improvement" in air quality in the Beijing–Tianjin–Hebei (BTH) region.[1] Nationwide, concentration of PM2.5 dropped an average of 38 percent to 45 micrograms per cubic meter ($\mu g/m^3$) in 2017. The number of prefecture-level cities[2] meeting the country's air quality standards increased from three in 2012 to ninety-nine in 2017.[3] Meanwhile, the annual percentage of days with "good air quality" (PM2.5 of 75 $\mu g/m^3$ or lower) in seventy-four cities increased from 61 percent to 73 percent on average, while the number of "heavily polluted days" (PM2.5 of 150 or higher) dropped from 8.7 percent to 3 percent.[4] Between 2013 and 2017, PM2.5 concentration dropped in the BTH region by 39 percent (beating the 25 percent target), in the Yangtze River Delta by 34 percent (versus a 20 percent target), and in the Pearl River Delta by 26 percent (against the 15 percent target).[5]

2017 government campaign against air pollution in Beijing and surrounding cities, coupled with favorable weather conditions, chalked up a significant success. In the last quarter of 2017, PM2.5 levels in the twenty-eight northern cities were down 33.1 percent compared with the same period a year earlier.[6] The annual average of PM2.5 concentration in Beijing dropped by 35 percent between 2013 and 2017, from 90 $\mu g/m^3$ to 58 $\mu g/m^3$.

By the end of 2017, China seemed to have accomplished all the major goals set in its 2013 action plan (see Figure 5.1). Based on data from 250 government air monitors throughout the country, Michael Greenstone

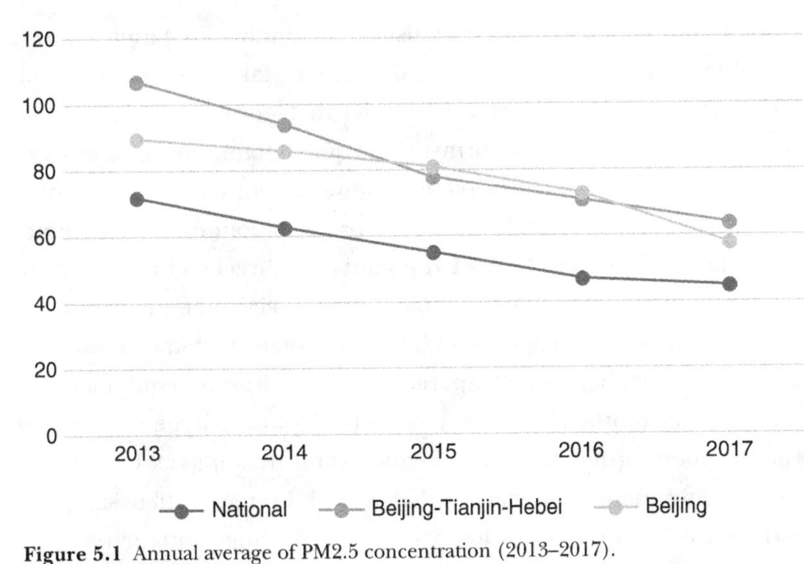

Figure 5.1 Annual average of PM2.5 concentration (2013–2017).
Source: Beijing Environmental Protection Bureau; Ministry of Ecology and Environment; *2017 nian zhongguo 365 ge chengshi lantian baoweizhan chengjidan* (Scorecard for Blue Sky Protection Campaign in 365 Chinese cities in 2017). www.drmj.org/2017–365-city-blue-sky-score.html

of the University of Chicago found that many of China's most densely populated cities recorded major declines in air pollution. This includes two of China's most polluted cities: Shijiazhuang and Baoding, which reduced their PM concentration by 39 percent and 38 percent, respectively. "Four years after declaring war on pollution, China is winning," he declared.[7]

Government anti-pollution efforts also have generated measurable health benefits. A study conducted by Tsinghua University found that national population-weighted PM2.5 reductions resulted in the drop of premature adult mortality from 1.22 million in 2013 to 1.1 million in 2015, or 9 percent of the attributable mortality abatements.[8] Greenpeace estimated that the improved air quality was associated with 160,000 fewer premature deaths across China in 2017.[9] Based on government data from 204 prefecture-level cities, Greenstone converted the reduced PM concentrations into their effect on life spans and concluded that Chinese people could expect to live 2.4 years longer, provided the decline in air pollution persists.[10]

But how reliable are these statistics? Beginning in January 2017, a number of provinces in China admitted to faking economic data, which led the public to suspect, rightly, that government air quality data are sometimes untrustworthy.[11] Air quality monitoring stations in China tend to be installed in parks or government compounds, which may not accurately reflect the actual exposure to pollution by surrounding residents. Moreover, the MEP does not have direct control over all of the monitoring stations, which enabled local officials, at times, to manipulate air quality data to appear to hit their targets. In some cities, local governments deliberately tampered with monitoring equipment by spraying water on the air quality sensors or deploying large air purifiers near the monitoring stations to reduce smog readings.[12] In extreme cases, environmental officials in Xi'an and Linfen stuffed sampling head devices with cotton to lower emission readings and/or removed surveillance tapes to cover up their fraud.[13] A joint study conducted by Greenpeace and Peking University found that the average daily reading of PM2.5 levels in Beijing in 2014 was 15 percent higher than the government statistics, and that air quality in the capital was twice as severe as the government reported.[14]

Fortunately, falsification of data appears to be on the wane. For one thing, PM2.5 is directly measurable and does not need to be recalculated or converted, providing less room for local officials to manipulate reporting.[15] Improvement in the quality of the data was indicated by a government analysis that identified twenty-three incidents of human tampering with air quality monitoring between November 2016 and July 2018, with only two occurring in the first six months of 2018.[16] The increased use of high-tech instruments, such as drones, scout robots, and remote sensing monitoring in environmental enforcement, also has made it easier to discover and discipline data fraud.[17]

Still, to determine as accurately as possible the improvement in air quality, I used data from the US Department of State's Mission China air quality monitoring program, which provides information on the changing level of air quality in five major Chinese cities: Beijing, Shenyang, Shanghai, Chengdu, and Guangzhou. The disadvantage is that air quality data are only provided in a single location for each city (i.e., the mission compound). But it gives us insight into the changing air quality in

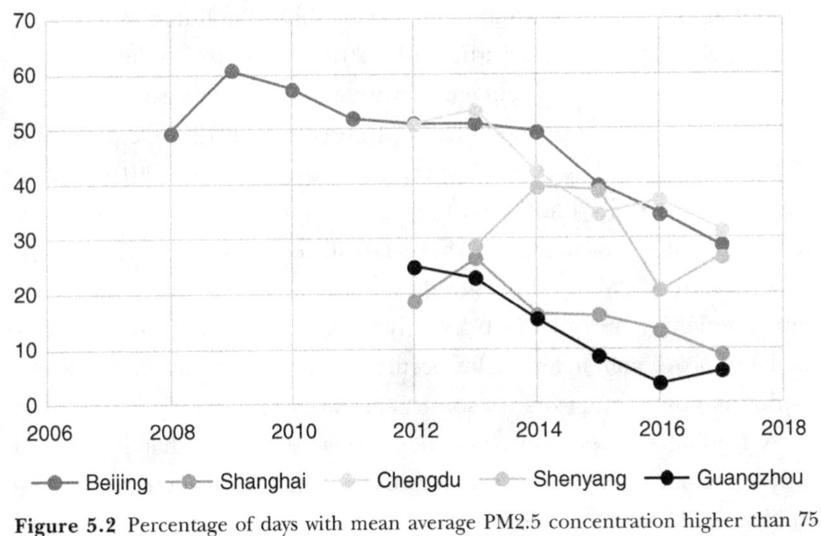

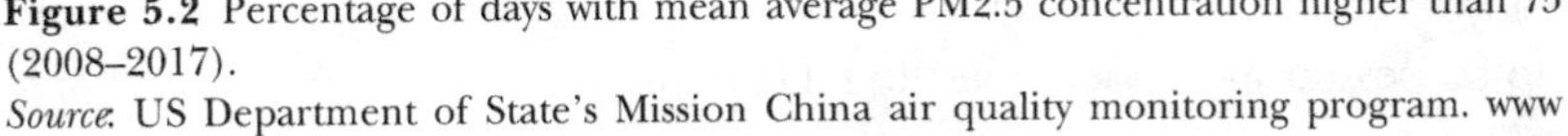

Figure 5.2 Percentage of days with mean average PM2.5 concentration higher than 75 (2008–2017).

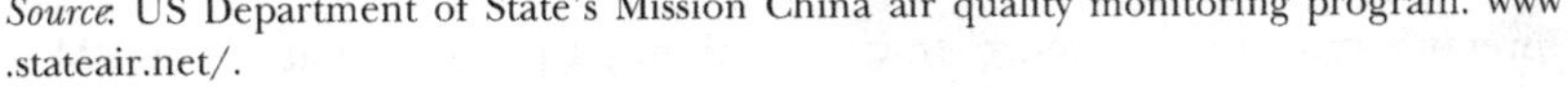

Source: US Department of State's Mission China air quality monitoring program. www.stateair.net/.

a reliable and trustworthy way. As Figure 5.2 shows, there has been an overall reduction in the percentage of polluted days (when PM2.5 average was higher than 75) since 2012. Guangzhou's drop was the largest, from 25 percent to 3.6 percent in 2016. Even Beijing, with the percentage of polluted days hovering around 50 percent during 2011–2014, enjoyed a decline from 49.6 percent in 2014 to 34.4 percent in 2016. Unfortunately, for 2017, the US Mission data only covers the first six months, so the effectiveness of the 2017 campaign cannot be fully assessed. The unusually high PM2.5 level in the first half of the year led to an increase in the percentage of polluted days for two cities in 2017 (Figure 5.2), but there is no denying that major improvements are happening.

THE LIMITS

Unlike smog in many other international cities, air pollution in China is seasonal. It typically gets worse with the arrival of winter in northern China due to heating emissions. If we look only at the PM2.5 level in the winter (November, December, and January), Beijing's experience with bad air quality barely improved from 2012 through 2016 (Figure 5.3). The percentage of winter days that were at least "slightly polluted" (PM2.5 >75 $\mu g/m^3$) actually increased during 2014–2016 despite the initial drop during 2012–2014. The percentage of winter days with "heavy pollution" (PM2.5 >150 $\mu g/m^3$) dropped from 31 percent in 2012 to 22 percent in 2014, but bounced back to 32 percent in the winter of 2016. Former Beijing mayor Cai Qi admitted that despite the drop in the number of heavily polluted days in the city, those days still contributed to 31.5 percent of the annual PM2.5 concentration.[18] As there has been an overall improvement in air quality on a yearly basis, one can conclude that most of the improvement in air quality occurred in other seasons. In fact, winter pollution was getting worse in some Chinese cities.

A leading Chinese environmental expert estimates that it will take Beijing municipality another ten to fifteen years to achieve an average annual PM2.5 concentration level below 35 $\mu g/m^3$.[19] He has good reason to be pessimistic. Indeed, in the BTH region, the concentration of PM2.5 precursors – any chemical that contributes to the formation of PM2.5 particles – remains ten times higher than in advanced industrial countries.[20] The Environmental Performance Index (EPI), produced jointly by Yale University and Columbia University in collaboration with

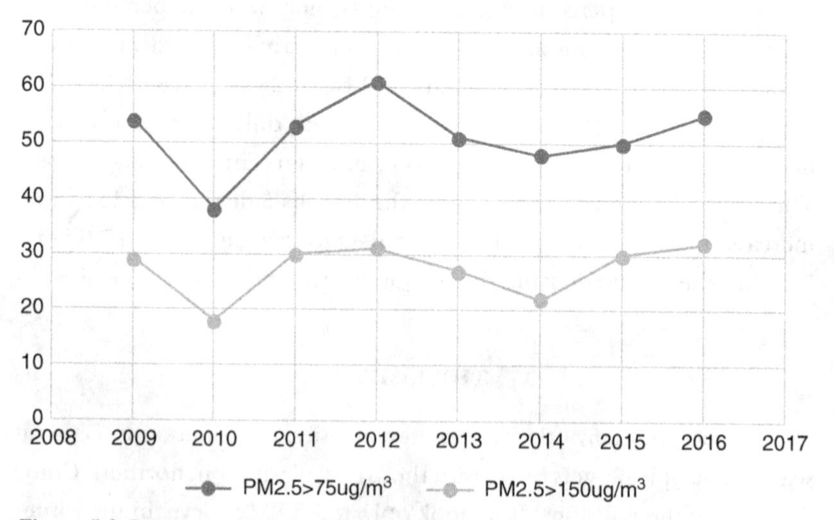

Figure 5.3 Percentage of winter days with bad air quality (Beijing, 2009–2016)
Source: US Department of State's Mission China air quality monitoring program. www.stateair.net/

the World Economic Forum, scores 180 countries on 24 environmental performance indicators. It ranked China 177 in air quality in 2018, trailing India, Bangladesh, and Nepal.[21] Estimations show that a significant drop in total emissions of main pollutants such as PM, SO_2, and NO_x will not come until around 2027.[22]

True, Beijing and the surrounding cities hit their 2017 air pollution control targets, which led to a further drop in the pollution level, but it seems to be mainly the result of cooperative weather conditions and a feverish last-minute anti-smog campaign. The progress was uneven nationwide. Between 2016 and 2017, while annual PM2.5 average in the BTH region dropped by 8.5 percent, PM2.5 levels rose in Heilongjiang, Anhui, Jiangxi, and Guangdong provinces by 10 percent, 7 percent, 4 percent, and 5 percent, respectively. That might explain why the overall drop in the PM2.5 concentration was only four percent nationwide during the same period.[23] Of the 388 cities at or above prefectural level, only a quarter of them hit their air quality improvement targets in 2017.[24] The analysis lends support to the argument that stricter and more centralized policy enforcement measures have not yet yielded significant effects in reducing overall pollution in China.[25]

More importantly, the fall in PM2.5 concentration has not translated into significant and sustained health benefits. The authors of the Tsinghua study admitted that the health benefits of reduced PM2.5 concentrations remain "limited" compared to the magnitude of air quality improvements, and mortality burdens induced by air pollution remained severe in China.[26] In the 2018 EPI, China ranked 167 out of 180 countries in environmental health and, together with India and Pakistan, is considered "the dominant source of disease and disability" in the data.[27]

Unfortunately, one-sided pursuit of PM2.5 reduction is problematic for improving public health outcomes in the long run. First, it does not help contain other pollutants that were largely untouched by the actions taken in China, such as ozone (O_3). Scientists have found that less PM combined with record industry output and hot weather creates a favorable environment for increased ozone formation.[28] Average ozone exposure in China rose by 17 percent during 2014–2017, causing an estimated 12,000 premature deaths annually.[29] This might still underestimate the health hazards of ozone exposure. According to the Global

Burden of Disease data, an additional 178,000 chronic respiratory disease-related deaths in China were attributable to ozone in 2017.[30]

Second, the focus on air quality improvement can distract attention and resources from prevention and control of other environmental health woes, such as water and soil pollution. By January 2018, China's central environmental inspectors reported that all but four provinces have either serious water pollution problems or they are not doing enough to tackle water pollution.[31] In addition, Chinese policymakers have yet to address the soil pollution issue with the same vigor applied to air pollution. Almost all the provinces dismissed soil contamination prevention and control with a few cursory lines in their 2017 government work reports.[32] The government still does not yet have a clear idea of the scale and scope of soil pollution in China, even with the 2014 survey data.[33]

Third, even if China manages to meet its goals on PM2.5 reduction, it will still fall well short of reaching the internationally acceptable level. The limits of the existing approach are demonstrated in a 2016 study by the Boston-based Health Effects Institute (HEI): in the most aggressive scenario, by 2030 PM2.5 concentration in China could drop to as low as 27 $\mu g/m^3$, a level still above the WHO guideline of annual mean PM2.5 of 10 $\mu g/m^3$.[34] As China's population ages and becomes more susceptible to pollution-associated diseases, the health benefits brought by improvements in air quality could be offset by a sizable aging population. A recent study by Chinese scientists estimated that PM2.5-related premature deaths would increase by more than 84,000 by 2020 and by 244,000 by 2030.[35]

Last but not least, the government campaign against air pollution, which targeted regions such as BTH, may have the unintended effect of outsourcing pollution from BTH to other parts of China. This can happen when pollution-intensive factories are relocated to regions with lower standards, making overall pollution worse. And even the gains for BTH are undermined to the extent that atmospheric transmission of outsourced pollutants boomerang back to the region.[36]

EVALUATING PROCEDURAL IMPROVEMENTS

It's evident that China's fight to control pollution has so far achieved only mixed results in meeting its tangible policy targets. If, however, one also sees benefit in inducing policy actors with diverse interests and preferences to tackle common challenges, the process itself may well be incrementally effective even if change is not immediate or substantial. This makes it necessary to adopt a procedural approach to assess the effects of policy – an approach that focuses on investigating whether a policy is pursued in ways consistent with the procedures called for. This alternative approach becomes particularly useful when assessing the effects of policy laid out in the three action plans – plans that include intervening steps on the way to final policy goals. Indeed, the central government began to conduct annual internal assessments of the progress of air pollution control as early as 2014, even though the final goals in the action plan were to expire at the end of 2017.[37] An examination of the follow-through on the stipulated steps and subgoals also enables us to get a glimpse of the "infrastructural power," or "the institutional capacity of a central state, despotic or not, to penetrate its territories and logistically implement decision."[38]

In the remainder of this chapter, I will examine the success and failure of the government in tracking the ongoing targets it set for achieving final policy goals, paying special attention to the following issue areas: reduction of toxic emissions, attacking water pollution, energy and industrial restructuring, policy coordination, public participation, and the use of market mechanisms. The government has highlighted these areas as essential to bring pollution under control.

REDUCTION OF TOXIC EMISSIONS. The government has set targets to cut a number of toxic emissions, including sulfur dioxide (SO_2) and nitrogen oxide (NO_x). The 2014 Government Work Report, for example, proposed to accomplish desulfurization in coal-fired power plants with a production capacity of 15 million kilowatts, denitrification in plants with a production capacity of 130 million kilowatts, and dust removal in those with a production capacity of 180 million kilowatts. It also proposed removing six million old high-emission vehicles from the nation's roads and providing high-quality diesel fuel for vehicles that meet new national standards. These goals were reported to have been overfulfilled by the end of the year.[39] Indeed, by the end of 2015, thermal power desulfurization and denitrification capacity ratio had reached 99 percent and 92 percent of the total installed capacity, respectively.[40] The concentration of major pollutants dropped further as a result of the large-scale ultra-low emission transformation work in 2016. Concentration of smoke dust, SO_2 and NO_x, was only 5 percent, 2.9 percent, and 4.5 percent of the 1997 level, respectively.[41] The emission limits for SO_2, NO_x, and smoke dust in coal-powered plants have reached the level of gas-powered plants and dropped below the state standards for effluent discharge by 83 percent, 50 percent, and 67 percent, respectively.[42] An analysis of major constituents of PM2.5 pollution in Beijing found that the share of sulfate dropped to below 10 percent, suggesting that government emissions control worked, at least in the short run.[43]

But is China pursuing the most cost-effective approach to improving health through pollution control? Each year, China consumes around four billion tons of coal. About half are used for power generation, and the other half for industrial and domestic use. In 2015, more than 70 percent of the iron and steel mills were found to be discharging excessive pollutants.[44] In Hebei province, each year 30 million tons of bulk coal were burned without taking any emission control measures, adding as much pollutant to the air as 1.5 billion tons of emissions from coal power plants.[45] Yet until 2016 the government did not pay serious attention to emission problems caused by the coal for industrial and domestic use. Instead, it applied aggressive desulfurization, denitrification, and dust removal measures to coal power plants. In doing so, the government provides power plants a subsidy of 2.7 cents per kilowatt-hour. A majority of the subsidies reportedly went to the pockets of major state power producers, such as Shenhua and Datang, as well as the several hundred newly established desulfurization and denitrification companies.[46] A State Council executive meeting in December 2015 decided to implement further ultra-low emission measures nationwide.[47] Under the new implementation plan, power plants that agreed to pursue ultra-low emission control measures would immediately receive an additional subsidy of 0.5 to 1 cent per kilowatt-hour. Doing so would cost an additional 100 billion *yuan* ($16.1 billion), but would not significantly contribute to smog reduction,[48] and could tilt the playing field in favor of building more coal power plants at the expense of new energy-saving and environment-friendly industries.[49]

Analysis also suggests a lack of effectiveness in controlling vehicle emissions, which release nitrogen oxides or NO_x. Nitrates still account for more than 30 percent of PM2.5 pollution.[50] Indeed, vehicle emissions increasingly are becoming the top source of PM2.5 in Beijing. Emissions from vehicles and other moving sources (e.g., construction machinery) contribute up to 45 percent of the total pollutants in the city, according to a study by Beijing Environmental Protection Bureau.[51] Nationwide, while coal emissions continue to account for most of the PM2.5 pollution, moving sources have contributed between 22 and 52 percent of PM2.5 concentration in Beijing, Shanghai, Guangzhou, and Shenzhen.[52]

Perhaps more important still, there is little indication that government emission reduction measures have led to any fundamental change in polluters' behavior. A 2015 study conducted by Greenpeace collected emission data from twelve power plants that claimed to have completed ultra-low emission transformation. It found that emissions in each plant exceeded the ultra-low emission limits.[53] It was reported that many of them turned on the denitration and desulfurization facilities during the day but shut them down at night, when they discharged more pollutants. In August 2017, central inspectors discovered that more than 55 percent of the 22,620 firms they visited had violated environmental regulations and laws.[54]

ATTACKING WATER POLLUTION. When examining the congruence between policy goals and outcomes, it is impossible to fully assess the effectiveness in water or soil pollution control because the two action plans (*shui shi tiao, tu shi tiao*) set year 2020 and beyond to accomplish their main policy goals. *Shui shi tiao,* for example, identified 2020 as the year to achieve "staged improvement" in water quality, including having over 70 percent of water in seven major river basins safe for drinking, fishing, and swimming. It also makes 2030 the year to eliminate black-odor rivers in cities and have 95 percent of drinking water sources in cities safe for human consumption.

Thus far, there have been promising signs of improving water quality. Between 2012 and 2016, the share of major rivers and lakes that met the water quality standards increased from 63.5 percent to 73.4 percent.[55] In addition, after five years of decline, groundwater quality improved in

2016, with the share falling under the "very bad" category lessening, although water unfit for consumption or swimming in lakes and reservoirs rose to 34 percent.[56] Yet progress in treating China's most polluted "black-odor rivers" has been slow. Between 2016 and 2017, the number of such rivers increased by 13 percent.[57] By June 2017, of the 2,100 black-odor rivers identified by the *shui shi tiao* for regulation and renovation, work had been completed on only 44.1 percent of them, suggesting that the timetable was slipping. In fact, fewer than half the rivers had been treated even though more than half of the designated time before the deadline had elapsed. In Hebei, Shanxi, Liaoning, and Anhui provinces, no action had been taken on 30 percent of the worst-polluted rivers.[58] Government ability to treat wastewater continues to lag behind. After the water pollution of the Weitan River in Shandong province was identified by central inspectors in 2017, the provincial government promised to fix the problem by June 2018. In March 2018, local officials began to throw chemicals into the river. The chemicals masked the pollution problem and allowed the project to pass the approval process four months later. The whole project cost 47 million *yuan* ($6.8 million). Yet in August, the water quality of the river began to deteriorate and returned to the worst level on the measurable scale.[59]

Controlling water pollution is of particular concern in the countryside, where nine billion tons of sewage are produced each year. Only 22 percent of that sewage has been treated, far lower than the treatment rate of over 90 percent in the cities. In Jiangsu province, for example, 146 of the 195 sewage treatment facilities are lying idle.[60] No wonder that NPC Chairman Li Zhanshu admitted in a speech delivered in August 2019 that "overall, our country's water eco-environment remains not optimistic."[61]

ENERGY AND INDUSTRIAL RESTRUCTURING. In the long run, emission control will not succeed unless China manages to shift away from its dependence on coal for energy and on heavy industry to drive its economy. Not surprisingly, government action plans against pollution seek to increase the share of clean energies such as hydro, wind, solar, natural gas, and nuclear power, while phasing out high pollution and high energy-consuming industries like cement and steel. Good progress has been made over the past six years. In 2014, when I visited my hometown, local government officials told me that they were experiencing an industrial upgrade: investments that would result in heavy pollution were no longer welcome.[62] This trend was matched at the national level by the shifts in the main drivers of China's economy from intensive steel and cement manufacturing to light industry, services, and transportation. In 2011, for the first time, the contribution to China's GDP from tertiary industry (the service sector) exceeded the contribution from secondary industry (construction and manufacturing). Six years later, secondary industry as a share of industrial production dropped further, to 40.5 percent, while the share of tertiary industry expanded to 51.6 percent.[63] China also has been making progress in eliminating excessive steel production capacity and curbing coal production. Steel output appeared to peak in 2014, and after that dropped by 2.19 percent in 2015 for the first time (Figure 5.4). Similarly, coal production in China apparently peaked in 2013 (Figure 5.5). The number of coal mines dropped by 35 percent during 2015–2017 to around 7,000 from 10,800.[64]

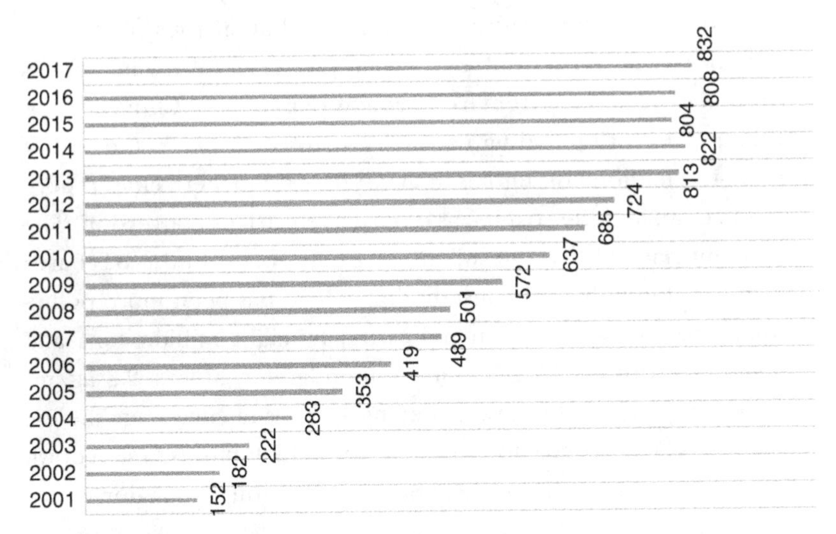

Figure 5.4 Steel output (2001–2017, in million tons).
Source: Dong Dengxin, "Zhongguo gangtie chanliang buru ling zengzhang shidai" (China's steel output enters zero-growth era); "2017 nian zhongguo cugang chanliang tongbi dazing 5.7%" (China's crude steel output saw an year-over-year increase of 5.7% in 2017), caixin wang, January 19, 2018. http://companies.caixin.com/2018-01-19/101200015.html.

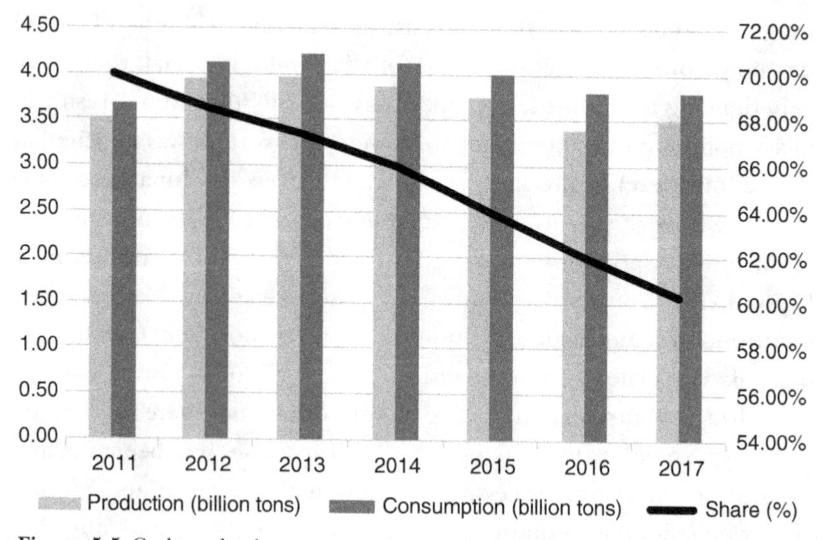

Figure 5.5 Coal production, consumption, and percentage share of the energy mix (2010–2017).
Source: Author's database.

The energy sector has been shifting as well. Clean energy sources contributed nearly 20 percent in 2016, up from 15.5 percent in 2013.[65] China's coal consumption dropped from 4.2 billion tons in 2013 to 3.8 billion tons in 2016, resulting in a drop in coal's share in China's energy mix to 62 percent from 67.4 percent (Figure 5.5). China accelerated the process of upgrading its energy sector in 2017. By the end of the year, 4.74 million households in the 28 largest northern cities reportedly had completed the coal-to-gas or coal-to-electricity conversion. The government claimed that the switch to cleaner energy in the BTH area had reduced the bulk coal consumption, which has been more polluting than coal consumption in industries, by eight million tons and resulted in the drop in PM2.5 concentration level by at least 2.3 $\mu g/m^3$ in the region.[66] In part because of Beijing's efforts to encourage the use of natural gas in lieu of coal for winter heating and its moves to shut down polluting smokestack factories, coal burning is no longer a major source of PM2.5 in the city.

Looking to diminish pollution from vehicles, China sold 777,000 electric cars in 2017, accounting for half of the cars that run on alternative fuel worldwide.[67] China's Minister of Science and Technology predicted that by 2020, new electric cars sold in China would reach two million and total electric car ownership would reach five million.[68]

Still, it is too early to conclude that there has been a fundamental, irreversible shift in the Chinese economy. China still accounts for more than 50 percent of the world's steel, cement, and flat glass production. Production and sales of traditional vehicles have set the pace for the world eight years in a row.[69] As Figure 5.4 shows, despite government efforts to cut excessive production capacity, steel production stopped declining in 2016, and actually rose 5.7 percent in 2017. A Chinese journalist found that steel production in Hebei increased in 2016 despite repeated government calls for eliminating excessive production capacity because "the cost for [managers of] steel mills to bribe local governments [to sustain overcapacity in the steel industry] is much lower than the cost of cutting overcapacity and investing in environmental protection."[70] Both coal production and consumption rose in 2017, raising the question of whether the production and consumption actually peaked in 2013 (Figure 5.5).

Weakness of policy integration, central-local preference gap, and lack of synergy between national development policies and environmental protection are rendering energy and industrial restructuring efforts less effective. While the central government sought to rein in the overbuilding of new coal-fired power plants and reduce the share of coal in the energy mix, it also unintentionally shifted authority over coal plant construction approval to provincial governments. With strong incentives to keep their local economies ticking along, provincial authorities seized the moment to approve new coal projects during 2014–2016. According to a report by the environmental group CoalSwarm, which used satellite images and official documents to track fossil fuel infrastructure, as of July 2018, 259 gigawatts (GW) of new coal-fired capacity were under development in China, equivalent to the total installed coal power capacity of the United States and enough to increase China's current coal power production (993 GW) by 25 percent.[71] Meanwhile, even if China hits its 2020 targets for electric car ownership, electric cars will still claim only 2 percent of the total number of cars on the road and an electric power sector still largely dependent on coal will only transfer emissions from vehicle tailpipes to utility smokestacks.[72]

POLICY COORDINATION. The government has made assiduous efforts to pursue an integrated government response in pollution control. Cooperation among the MEP, the National Development and Reform Commission (NDRC), the Ministry of Finance (MOF), and the Ministry of Commerce (MOFCOM) enabled the central government to release more than twenty supporting measures for implementing the 2013 air pollution action plan, including offering electricity price subsidies for ultra-low emissions, raising the pollutant charge standards, levying pollution charges for volatile organic compounds (VOCs), upgrading refined oil products, and strengthening coal quality management.[73] At the regional level, leading small groups have been established to harmonize planning, surveillance, policy enforcement, and emergency response. The smog-prone region around Beijing is a prime example. As early as September 2013, with support from the MEP and the NDRC, a coordination mechanism was set up to cover six provinces and autonomous regions in northern China. In 2016, Beijing municipality earmarked 500 million *yuan* ($72 million) to support the use of clean energies in two cities in neighboring Hebei province.[74] The same year, MEP kicked off regional emergency response plans to address heavy smog outbreaks in the BTH region. It implemented a unified standard for pollution alerts across the BTH region in February, followed by a document requiring ten cities across the region and in nearby Henan and Shandong provinces to take joint action.[75] On November 4, 2016, prior to the arrival of heavy smog, the governments of Beijing, Tianjin, and Hebei coordinated an orange smog alert, the second highest danger level. Anticipating the smog was on its way to Beijing, governments in the surrounding areas undertook emergency emission control measures. This experience prompted the government to move toward institutionalizing cross-regional environmental governance in northern China by setting up a joint environmental protection agency.[76] The payoff was evident in the 2017 “2+26” plan on prevention and control of air pollution, which was issued jointly by the MEP, the NDRC, the MOF, the NEA, and the provincial or municipal governments of Beijing, Tianjin, Hebei, Henan, Shandong, and Shanxi. The concerted efforts reportedly contributed to a 20 percent reduction in main pollutants in the air.[77]

Yet, often times, effective interagency and region-based coordination is only achieved during the campaign stage. As far as environmental health is concerned, the two functional areas (environment and health) remain segregated, and institutionalized cooperation between departments is still rare. The 2007 action plan on environmental health sought to mitigate this problem, but failed to address the issues of functional overlapping and the regulatory "blindspot" between the two ministries, leading to bureaucratic infighting and buck-passing in implementation.[78] The problem lingers on. According to an assessment made by the MEP in 2017, "for a long time there are no institutional requirements for national environmental health work; there are no explicit requirements for functional differentiation and coordination models between central and local governments, or between the MEP and health departments; and local environmental protection departments are still not sure whether they should conduct environmental health work and how they should carry out environmental health duties."[79] At the county level, for example, there is little cooperation between the environmental monitoring stations and the local CDC in assessing the health impact of local environmental problems.[80]

Both the updated environmental protection law of 2015 and the air pollution control law of 2016 have articles on environmental health, but neither law clearly delineates the responsibilities of the National Health and Family Planning Commission (NHFPC) or the MEP. As a result, the goals of environmental protection are not aligned with those of health, and public health research findings are not integrated into environmental policymaking. Additionally, while the three action plans released after 2013 aimed to improve public health through reductions in pollution, health departments essentially were excluded from the policymaking.[81] Similarly, health bureaucracy is marginalized in the implementation process. In the action plan for water pollution prevention and control, the NHFPC is not listed as a lead agency in any issue areas related to environmental health; it is only included as a "participating agency" in four of the eight issue areas.

Due to the lack of coordination between the two ministries, thus far there has been no MEP–NHFPC joint survey research on the health effects of environmental pollution, even though environmental health experts from China CDC have been invited to participate in MEP-sponsored environmental studies. There is no data-sharing mechanism between the two departments, either. Environmental health researchers at China CDC reportedly had to copy information from the websites of the National Environment Monitoring Station (NEMS) on a daily basis – the data was only in PDF format and was not downloadable. When requesting air quality monitoring data for January 2013 from the MEP, all the CDC researchers got was the NEMS website.[82] Data sharing is particularly a problem at the subnational level.[83] When health departments ask for data from environmental departments, the latter often refuse to share the information, claiming that environmental data needs to be kept secret. A circular issued by Shaanxi's environmental protection department, for example, made it clear that any third party that wants to use the province's soil pollution survey data needs to obtain prior authorization from the head of the department.[84]

The coordination problem makes it difficult to assess the health effects of environmental change. For example, the MEP's air quality monitoring sites usually are set up 10 to 15 meters high, even though the human respiratory zone is only 1.5 meters high. PM2.5 concentrations obtained this way don't reflect the actual level of human exposure to ambient PM2.5. Such concerns prompted the NHFPC to build its own projects to monitor the health impacts of air pollution. By January 2017, the NHFPC monitoring system included 125 monitoring stations, covering 60 cities in 31 provinces.[85] In contrast to MEP's reports, NHFPC monitoring data found only a "somewhat" improvement in air quality between 2013 and 2017, which did not lead to significant improvement in people's health status.[86]

PUBLIC PARTICIPATION. Despite the tightening of social controls, the central government's stated support for public participation in the environmental policy process has contributed to the proliferation of environmental NGOs in China. It is estimated that there are 8,000 environmental NGOs in the country today, compared to just 9 in 1994.[87] Some environmental NGOs have played a prominent role in promoting environmental health. Led by Isabel Hilton, London-based *chinadialogue* is a bilingual website that fosters a common understanding of urgent environmental challenges in China. They publish and translate articles, blog posts, and reports written by authors from inside and outside China, focusing on topics that cover pollution, health, and food.[88] In 2007, environmental activist and journalist Feng Yongfeng founded Nature University, an NGO that offers training and resources on environmental protection. Convinced that what the government feared most was socialization of environmental problems, Feng encouraged the use of multiple means – petitions, seminars, and surveys – to "continuously harass" the government and force it to take steps against pollution.[89]

Since 2010, Greenpeace East Asia has been actively promoting environmental health by investigating, documenting, and exposing industrial pollution in China. Much of its work highlights the exposure of the Chinese people to toxic chemicals, such as the hazards of coal ash and phthalates used to soften plastics used in children's toys. Food safety has consistently been a key focus. The NGO has exposed the presence of nonylphenols in wild Yangtze River fish, found banned pesticides in the products marketed by Chinese tea companies, and brought attention to the cadmium contamination in rice produced in Hunan province. Since 2012, it has taken on air quality issues in China. In December 2012, it conducted a study with Peking University's School of Public Health to estimate the premature deaths and economic losses caused by high levels of PM2.5 pollution. This was followed by another study in February 2015, which showed how high PM2.5 concentration was contributing to an increased urban mortality rate. Greenpeace sometimes takes action in reaction to the problems exposed. Following the investigation of heavy metal pollution in Hunan, it brought 500 kilograms of tainted rice directly to the provincial government. In October 2014, they projected a message on the Beijing Drum Bell Tower asking for "BLUE SKY NOW!" Such efforts drew attention to the health effects of air, water, and soil pollution, which prompted central and local government officials to respond in a swifter and more effective manner.[90]

Compared to the self-reliant, sometimes high-profile approach of Greenpeace, Ma Jun's IPE developed a more cooperative working relationship with the government in conducting evidence-based research and advocacy. Ma, a journalist turned environmentalist, founded the NGO and civilian think tank in 2006, with the belief that increasing access to environmental information could bring public pressure to bear on government actors and industrial firms. To that end, IPE built an online database and a map of real-time emissions that enables the public to check on air and water quality, locate sources of pollution, and monitor emissions from polluting firms. Unlike Greenpeace, IPE does not collect pollution information independently. Rather, it collates and analyzes data based on publicly available governmental information (mostly from the MEP and the Ministry of Water Resources). This, according to Ma, also forces IPE to deal with government agencies. Ma denies that IPE are "fighters" for environmental protection. In a phone interview, Ma repeatedly talked about "constructive interaction" with central and local governments in uncovering violations of environmental laws and regulations. "We take a multi-stakeholder approach," he said.[91] A non-confrontational approach has allowed IPE and other NGOs, including Friends of Nature, to participate in the making of environmental policy.[92]

In addition to raising awareness about environmental health issues, environmental NGOs in China have been expected to play a prominent role in using environmental public interest litigation (EPIL) to compel firms to comply with environmental laws. In 1998, two professors at China University of Political Science and Law co-founded the Center for Legal Assistance to Pollution Victims, which since then has brought numerous cases seeking redress for pollution victims. The new Environmental Protection Law, enacted in 2015, empowers NGOs to bring public interest lawsuits against polluters, which promises to create more space for public participation in China's environmental protection.

Still, China's environmental NGOs are yet to develop into a significant force to be reckoned with. In 2015, the budget of all NGOs combined was less than 100 million *yuan* ($16 million).[93] IPE, one of the largest environmental NGOs in China, had no more than sixteen staff members. Of all the environmental NGOs, fewer than 500 have the ability to take public action. An NGO leader told me that a majority of the environmental NGOs were engaged in raising awareness, but their impact was "not comparable to a haze."[94]

Only four NGOs lodged EPIL suits between 2007 and 2013.[95] With the new Environmental Protection Law in effect, the anticipated significant increase in the number of EPIL cases initiated by NGOs did not occur. While around 700 environmental NGOs were granted the right to initiate EPIL cases, only 9 of them – including Friends of Nature, the All-China Environment Federation, and the China Biodiversity Conservation and Green Development Foundation (CBCGDF) – filed 37 such cases in 2015.[96] The situation did not change in 2016. Even though the number of EPIL cases more than doubled over 2015, they accounted for only 8.3 percent of the total number of environmental pollution liability dispute cases (1,765).[97]

To the dismay of environmental activists, most of the EPIL lawsuits have been initiated thus far by government departments and local procuratorates.[98] In July 2015, the NPC Standing Committee authorized the Supreme People's Procuratorate, the country's highest agency responsible for both prosecution and investigation, to launch pilot EPIL programs in thirteen provincial units. By the end of 2016, 495 EPIL cases – most of which environmental ones – had been initiated by local procuratorates in these provinces.[99] Nationwide, the number of cases initiated by procuratorates and processed by courts increased by one third between 2017 and 2018, from 1,304 to 1,737.[100]

Public interest litigations are limited for several reasons. First, while the new Environmental Protection Law is heralded by environmentalists as a breakthrough, it did not level the playing field to allow effective participation by NGOs. As I noted elsewhere, even though the post-Mao reform dynamics have expanded space for health-related charity organizations, the state continues to dominate the non-state sector, which has negatively affected the registration, financing, operation, and expansion of NGOs and private foundations.[101]

Second, environmental NGOs tend to be small and lack the means to specialize in both the environment and the law, not to mention the need to defray high litigation costs. The reluctance of local governments and enterprises to disclose information on the environment only increases the cost of investigation and evidence collection. A typical case initiated by an environmental NGO would cost 200,000 to 300,000 *yuan* ($30,200 to $45,300).[102] In April 2016, Friends of Nature and CBCGDF took three chemical manufacturers to court for soil pollution in Changzhou, Jiangsu province. In January 2017, the local court ruled against the plaintiffs, forcing the two NGOs to pay 1.89 million *yuan* ($270,000) in court costs.[103]

Third, local governments, motivated by the need to sustain economic growth and reduce unemployment, often choose to connive in local firms' pollution behavior. Because most of China's environmental NGOs are based inside China, they are subject to interference by local governments in filing EPIL cases. Lawyers and activists hoping to bring cases often are harassed by police at the behest of the local authorities. The government not only decides which NGOs can file environment-related public interest lawsuits, but also controls the decisions of judges. As a result, it is extremely difficult to win an EPIL case against major local taxpayers unless the latter have committed other, more serious, crimes.

Fourth, the new law empowered NGOs to file EPIL lawsuits against polluting firms only, not government entities. In December 2016, Friends of Nature filed two administrative public interest cases against Nujiang Environmental Protection Bureau in Yunnan province, for violating the new law by approving the environmental impact assessment of a regional plant making potassium perchlorate, a volatile chemical often used as a solid rocket propellant. Production of potassium perchlorate generates dangerous by-products, including the carcinogenic hexavalent chromium. The local court initially accepted the cases in January 2017, but then dismissed them within five months, indicating the government is above the law it created.[104]

The marginal role of NGOs in filing EPIL cases is emblematic of the increasingly narrow public participation in environmental governance. While the government officially acknowledges the role of the public in the environmental policy process, it continues to restrict, even deny, people's rights to express themselves peacefully. In December 2016, Chengdu police broke up a small group of people wearing masks at an anti-pollution protest.[105] Today, even social media discussions on environmental health issues are monitored closely and government censors will remove any information they consider sensitive.

The authoritarian approach is clearly evidenced in the government campaign to promote trash recycling. Like the launch of the draconian one-child policy in the early 1980s, rollout of the compulsory garbage sorting program in Shanghai was not based on environmental awareness and enthusiasm from the bottom but imposed directly from the top. This "eco-dictatorship,"[106] as a Chinese scholar called it, was also demonstrated in the implementation stage. Research shows that community-level measures such as engaged volunteers are more effective than government introduced measures like posters and points-based systems in nurturing residents' new garbage-sorting habits.[107] The failure to engage the public also caused a great deal of confusion on the new rules.

MARKET MECHANISMS. Given the enormous scale of China's environmental health challenges, huge investments are needed to address air, water, and soil pollution. It is now estimated that to clean up pollution in China, at least 3 percent of GDP needs to be spent on environmental protection.[108]While public spending has increased sharply, the investment on environmental cleanup of 880 billion *yuan* in 2015 still accounted for only 1.3 percent of GDP.[109]

As China's economy faces increasing uncertainties and as local government debts continue to rise,[110] the government alone may not be able to sustain the spending required. A State Council document issued in October 2014 imposed constraints on local governments' ability to issue debt for public financing, exacerbating budgetary concerns. As a result, efforts were made to promote public–private partnerships (PPP) – long-term, renewable contracts between private entities and government agencies in which the former provides a public asset or service, bearing significant risk and management responsibility. In April 2015, the State Council adopted the draft "Administrative Measures on Concession of Infrastructure and Public Utilities Projects," which aims to encourage private sector participation in infrastructure projects. The same month, the MOF and the MEP formally introduced the PPP model to environmental protection. This was followed by the unveiling of a circular jointly issued by four central ministries calling for comprehensively implementing PPP in government-involved sewage and garbage disposal projects in July 2017. The government initiatives appear to be paying off. Among the more than one thousand PPP projects unveiled by the NDRC in 2015, a third were related to environmental protection, with investments amounting to $26.8 billion.[111]

Some of the projects were quite successful. During 2015–2018, CRRC Corporation invested one billion *yuan* ($160 million) in the water pollution control project in Changshu of east Jiangsu province. The PPP enabled the city to access advanced water treatment technologies to ensure the collection and treatment of domestic sewage of more than 21,000 households in 930 villages, and to have the treated water safe for farm irrigation and afforestation.[112] By the end of 2017, total revenue from environmental industries had reached 1.35 trillion *yuan*, 17.4 percent higher than the 2016 level. The number of listed companies dedicated to environmental protection, such as Zhongjin Environment Co. and Infore Environment Technology Group, also increased, from 45 in 2012 to 80 in 2016.[113]

In recognition of the limits of command-and-control measures,[114] the Chinese government also has employed tradable pollution rights as a market instrument. As early as April 2001, China's State Environmental Protection Administration (SEPA) and the New York-based Environmental Defense Fund jointly launched a project to promote the implementation of China's SO_2 emission trading policies. Emulating US market-based pollution control measures intended to promote cleanup as cheaply and efficiently as possible, China introduced fees and tradable pollution permits for industrial firms. In 2007, Jiaxing city of Zhejiang province set up the first pollution rights trading (PRT) center. Since then, twenty-eight provinces have set up trading platforms that allow industrial firms to buy and sell rights to emit pollutants such as SO_2 and NO_x.

The system constitutes two markets. In the primary market, industrial firms are required to pay for pollution rights through auction or government-directed quotas; firms then can trade permits in the secondary market, which allows firms that overproduce pollution to purchase additional permits from less polluting companies to "offset" their extra emissions. The government issued guidelines on the paid use and trading of pollution rights in August 2014 and interim measures for managing revenue from emission rights transfer in July 2015. In December 2016, the 13th Five-Year Plan for Ecological and Environmental Protection put forward the goal to establish cross-regional PRT market. By August 2018, receipts from the paid use and transfer of pollution rights amounted to $2.9 billion, including $1.8 billion from the primary market and $1.1 billion from the secondary market.[115]

Unfortunately, an underdeveloped market society and a weak environmental regulation regime have restricted the effective use of market mechanisms in environmental protection. Not only do PPP contract violations remain common, but private entities often are in a weak position to deal with government actors due to an absence of laws and regulations to ensure equal status in consultation and negotiation processes. In fact, institutional insufficiencies only have emboldened excessive government interference in PPP-related project setting, assessment, and decision-making. Many local governments still view PPPs as a financing tool and overlook the private sector's potential role as long-term, responsible partners. In the period from 2007–2015, the government claimed the lion's share in financing the 316 soil remediation projects nationwide.[116] The situation has not changed much since. In the face of the difficulties in raising capital, many environmental protection companies are shying away from PPP projects.[117]

Progress also has been limited in pollution rights trading. A review of China's PRT market by two Chinese scholars concludes that the market remains fragmented, trading information remains nontransparent, and a nation-wide trading system for pollution permits is still in the exploratory stage.[118] Of the twelve provinces authorized by the central government to experiment with PRT, two thirds of them have performed poorly, and only seven have piloted trading in the secondary market.[119] Since its debut in May 2008, the China Beijing Environment Exchange has not executed a single trade in pollution rights. Firms do not have strong incentives to go to the trading platforms in part because of the lax enforcement of environmental laws – it makes no sense to seek permits if the firms can discharge pollutants without the risk of being caught. The existing regulatory framework also allows a majority of local industrial firms to obtain a pollution-rights quota for no cost as long as they pass environmental impact assessments. The incentives for relying on market forces are curtailed further by the absence of a standardized, nationwide price-setting or regulation regime for polluting rights. Rather than playing the role only as a regulator, the government itself is active in cutting deals. Indeed, most of the deals are not genuine market-based pollution rights trading but are completed with the intervention of local environmental protection agencies, which deny the role of price mechanisms.[120]

Levying an environmental tax has further compressed the market – now any industrial firm purchasing pollution rights also must pay the tax.[121]

PERVERSE INCENTIVES AND UNINTENDED OUTCOMES

In a country where cost benefit analysis is rarely conducted, emission reduction measures are often implemented without seriously accounting for their unintended effects. As early as 2007, in response to the outbreak of the blue-green algae that choked Lake Tai, the government closed down 4,400 chemical industrial firms in the region. Some firms were then relocated to Anhui province. As sources of pollution moved to the upper reaches of the Lake Tai Basin, their threat to the environment loomed larger.[122] Similarly, in fighting air pollution in China, efforts to promote the rapid expansion of coal gasification – one of the highly water-intensive forms of energy production – have exacerbated China's water shortage problem.[123] Other emission control initiatives, especially those that aim at scrubbing coal-related pollutants or reducing coal use, may have concomitant benefits in helping China reach its near-term CO_2 reduction goals while improving air quality and environmental health. Nevertheless, as Valerie J. Karplus of MIT has pointed out, once those low-cost measures to reduce coal are exhausted, the installation of equipment to effect end-of-pipe pollution control will involve the use of carbon-intensive energy, which actually can increase CO_2 emissions and hinder the progress toward realizing China's pledge to reach "peak carbon" by 2030.[124]

To achieve continuous improvement in environmental health, it is essential to internalize and routinize pollution control measures, just as China did with population control.[125] The problem is that local governments prefer to handle the measures in a perfunctory manner, while environmental protection agencies do not have sufficient regulatory capacity to deal with the country's innumerable sources of pollution and widespread violations of environmental regulations.[126] Crash campaigns are still the preferred tool for policy implementation in China's pollution control. By setting a clear goal with a measurable outcome and by mobilizing the society and bureaucracy toward meeting the designated goal, campaigns help promote interagency collaboration while overriding fiscal constraints and bureaucratic inertia.

But campaign-style enforcement is ultimately an ineffective solution to China's environmental health challenges. It typically focuses on achieving short-term results. So even if it fulfills the campaign objectives, the results are costly to attain and difficult to sustain. Interdepartmental coordination can be realized, but it lasts only as long as the campaign goes on. Campaign-style enforcement tends to rely heavily on administrative fiat in lieu of public participation and market mechanisms, fostering a non-scientific and heavy-handed approach in policy implementation. The overreliance on an overbearing mobilization hinders the growth of stable, law-based enforcement mechanisms that are essential for long-time policy effectiveness. One-sided pursuit of quantifiable targets may be the only way to cut through the red tape inherent in China's complex and multi-layered bureaucracy, but it often leads to unintended and undesirable outcomes that undermine the achievement of other policy goals.

In Henan province, for example, environmental protection officials complained that the need to support the MEP-launched environmental inspection campaigns increased their workload by three to five times and that they could no longer attend to the day-to-day business of their organization. Also, since the inspection teams typically only stay in one location for two weeks, it was next to impossible to see any fundamental change in local bureaucrats' policy behavior in such a short period of time.[127] Very often, the MEP did the inspection but there were no follow-up rectification measures to sustain the change. In Tangshan city, Hebei province, the inspection team made a second "looking back" visit to thirty-nine places they had visited ten days earlier and found that the problems identified in all of them were left unaddressed.[128]

PARADE BLUE. The phenomenon of "APEC blue" or "parade blue" – temporary fixes to create cleaner air – epitomizes the process. During high-profile and politically sensitive time periods, the central and local governments are tempted to impose a series of strict emission control measures – closing factories, restricting traffic, halting construction – to demonstrate to visitors that they can get air pollution under control. This happened during the 2008 Beijing Olympics, the 2014 APEC summit, the 2015 parade celebrating victory over Japan in World War II, and the 2016 G20 Summit in Hangzhou.

During the APEC summit (November 1–12, 2014), air pollutants in Beijing reportedly fell to their lowest levels in five years.[129] But the cost of achieving APEC blue was high.[130] According to a study conducted by Peking University, the initial drop in PM2.5 concentration was not impressive when government emission control measures were pursued only in Beijing and some surrounding cities in Hebei province. The impact became truly significant only after almost all smog-generating economic activities in northern China (which includes part of Shandong, Shanxi, and Inner Mongolia in addition to Beijing and Hebei) were halted for one week (November 6–12). Even so, the average PM2.5 concentration during the week was 51.5 $\mu g/m^3$, more than twice the maximum average exposure in a 24-hour period recommended by the WHO – 25 $\mu g/m^3$.[131]

And even when CCP leaders try to temporarily reduce visible pollution, they are not always successful. In late September 2019, just days before the huge parade celebrating the 70th anniversary of the founding of the People's Republic of China, heavy smog swept over Beijing, prompting the city to issue its first orange smog alert of the fall. By that time, the government had already put in place measures for emission control based on its usual songbook for parade blue.[132] Like many China watchers, I expected those measures would result in clear skies in Beijing on October 1. To my surprise, at 10:00 a.m. of the National Day, when Premier Li Keqiang announced the beginning of the celebration, the concentration of PM2.5 reached 155, a "very unhealthy" level. And while President Xi was showing off China's military might, promising that "no force can shake the status of this great nation," a cloak of smog became a fly in the ointment, rendering fighter jets and helicopter gunships roaring past Tiananmen almost invisible.

Additional problems of the campaign are identified by a three-year analysis conducted by a team of Chinese researchers who sampled air quality in 189 cities before and after the local parliament and advisory body meetings (the "two sessions" or *lianghui*) during December 2013 and March 2016.[133] The analysis found significant improvement in air quality during the *lianghui* period, with AQI dropping by an average of 5.7 percent. Yet the same analysis also highlights the limits of single-minded emphasis on PM reduction in the cadre performance evaluation system: significant improvement only occurred for indicators that are used as a yardstick to measure local government performance in environmental protection, such as PM2.5 and PM10. There was no significant impact on other pollutants, such as ozone (O_3). The O_3 concentration actually increased during and after the *lianghui* period (Figure 5.6).

Central leaders appeared to be aware of the downside of the politically charged PM2.5-centered campaigns. During the APEC summit in 2014, President Xi made the following remarks:

> Some people say, the current blue sky in Beijing is "APEC blue," beautiful but transitional, and will disappear after the meeting. I hope and believe the APEC blue can be sustained through assiduous efforts.[134]

Much to the chagrin of President Xi, the smog returned as factories resumed production and traffic controls were lifted after the summit. The same pattern can be identified in the three-year analysis of 189

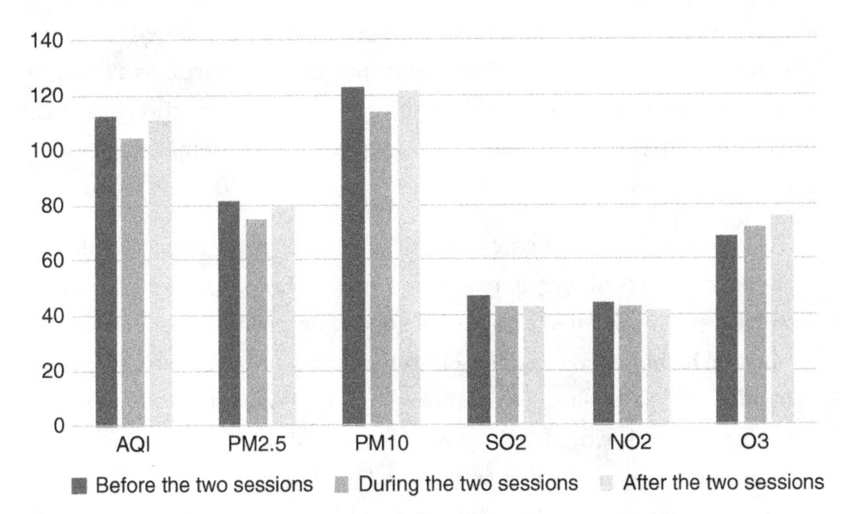

Figure 5.6 Pollution level before and after the "Two Sessions" in 189 cities
Source: Shi Qingling, Guo Feng, Chen Shiyi, "Wumai zhili zhong de zhengzhi xing lantian" (Political blue sky in smog control), *Zhongguo gongye jingji* (China Industrial Economics), no. 5, 2016.

Chinese cities. As Figure 5.6 shows, the end of the "two sessions" was followed by a rapid rise in pollution, with AQI, PM2.5, and PM10 quickly approaching the level prior to the two sessions. Such a "retaliatory rebound," also evident in the rapid recovery of PM2.5 in early 2017, shows the fragility of China's air quality improvement.

SHORT-TERM GAINS VS. LONG-TERM IMPROVEMENT. The 2016–2017 pollution control campaigns may have met their targets, but they also appear to have created a kind of destructive Frankenstein within the Chinese bureaucracy. Previously, inaction, foot dragging, and lax enforcement characterized local government officials' attitudes toward enforcing environmental protection laws and regulations over local industrial enterprises. Since the central government began environmental protection inspection campaigns in 2016, such "local protectionism" has shifted to "official protectionism." Due to strong pressures from above, fulfilling short-term targets has become the sole goal for local implementers caring about their personal careers. The problem is that they do not have the regulatory capacity to identify all the polluters. Yet to avoid being found guilty of shielding polluting factories and to meet the sometimes questionable goals quickly, local officials with an eye on career advancement – or simply fearful of being sacked – are inclined to undertake nonscientific, heavy-handed, and unpopular measures. As a Chinese environmental scholar observed, under this official protectionism, local officials tend to prefer a cookie-cutter approach in implementing central directives:

> They do not approve the permits [for environmental protection] that should be approved, they impose heavy fines on firms that should be hit with lower fines, and they close all firms on heavily polluted days no matter whether or not they meet the environmental standards. Such simple-minded and heavy-handed measures not only hurt local economic growth but also undermine the sustainability of environmental protection work. ... Such behaviors are found more in the central and western regions than in the eastern region.[135]

In a deadline-driven, high-pressure campaign environment, policy implementers at the micro level may know local conditions better than their superiors, but they have neither the incentives nor the flexibility to pursue more effective policy options. The central issue of the cookie-cutter approach is its reliance not on laws and environmental protection standards to address pollution but on minatory administrative fiat. Even some industrial enterprises that met advanced international environmental standards were ordered to shut down or at least have a portion of their production capacity closed down.[136]

At the same time, the indiscriminate closing of polluting industrial firms (most of them small and medium private enterprises) soon had secondary effects. One of the objectives of the 2017 clean-air campaign was to regulate and remove businesses considered "scattered, chaotic and dirty" (*san luan wu*). After closing down many industrial polluters, local regulators began to target small local private businesses, including restaurants, barbecue stalls, and car wash shops, causing a great deal of inconvenience in local people's lives.[137] In Dali, Yunnan province, the city government closed all 1,800 restaurants and hotels by Erhai in order to improve the lake's water quality.[138] Local leaders in Zhengzhou, the capital of Henan province, initially identified 539 *san luan wu* businesses. When they learned that they would be penalized if central inspectors found any additional firms not on the list that could be considered in violation of pollution rules, however, they expanded their lists to include many very small businesses such as auto repair shops or stalls making steamed buns. Within three months, the number surpassed 10,000, putting at risk mom-and-pop operations that cause very little pollution. This *Kafkaesque* policy implementation also was found in other localities, including Beijing, where a large number of supermarkets, department stores, and clothing stores were on the list for "clean up." By the end of June 2017, the BTH region reported 176,000 such firms in need of "clean up," three times more than the previously identified number.[139] Those firms that were unable to meet the emission standards were scheduled for closure by September.

Soon it became clear that the campaign against pollution was disrupting an already slowing economy. From July to September 2018, China's economy experienced the slowest quarter since 2009, with industrial profits slowing for the fifth consecutive month in September.[140] The flagging economy prompted the Ministry of Ecology and Environment (MEE), the successor to the MEP, to end the cookie-cutter approach in enforcing environmental protection. Instead of forcing factories to shut down whether they had implemented pollution controls or not, a MEE meeting in late October 2018 stressed the need to protect legitimate business operations. The Central Economic Work Conference, held two months later, reiterated the importance of "overall planning and all-round coordination" to "avoid oversimplified and crude measures" and "enhance service consciousness [in dealing with industrial firms]."[141]

Nonetheless, in localities facing mounting pressure to significantly and swiftly improve air quality, the central government directives fell on deaf ears. In August 2019, Shandong provincial authorities held an EST with officials of Linyi, and the city responded by threatening to remove local party chiefs from office if their townships were ranked among the bottom-two in air quality. The new measure drove officials of Lanshan district, which houses the country's largest board production base, to force over 400 board factories to suspend production whether the firms met environmental protection requirements or not.[142]

Campaign-style policy implementation is exacerbated by the rise of the bandwagon polity, which encourages local implementers to act in a way that is more Catholic than the Pope – overshooting with enforcement to demonstrate their enthusiasm for the policy aggressively promoted by Xi and avoid being accused of foot dragging. During the APEC summit, China suspended operation of nearly 10,000 firms, curtailed production of 3,900 firms, and halted work of 40,000 construction sites, which was 3.6 times, 2.1 times, and 7.6 times, respectively, of what was required under the plan.[143]

WINTER OF DISCONTENT. And then there were the unintended consequences of the campaign to convert coal-generated heating to gas or electric heating. A MEP survey in December 2017 identified more than 4.74 million households in the BTH region that had completed the conversion, of which only 17 percent were done in 2016. That means 83 percent of the households completed the conversion during the 2017 campaign. Interestingly, 795,000 households, or nearly 17 percent of the households that completed the conversion, had not been slated to do so. Furthermore, in addition to the twenty-eight cities prioritized for completing the conversion, another thirty cities in Shandong, Shanxi,

Henan, and Hebei went out of their way to convert an additional 1.5 million households (Figure 5.7).[144] A forerunner in this bandwagon polity was Hebei province. While the government plan was to convert 1.8 million homes in the province to clean energy by the end of October 2017, the implementation went so well there that more than 2.5 million homes ended up switching to gas or electricity.[145] Shanxi province followed the same "overshooting" pattern.[146]

One problem, however, was that some rural households were reportedly ordered to dismantle their coal-fired furnaces even before gas pipelines were laid to their houses.[147] An even bigger issue was that the haste to eliminate dirty coal-burning household furnaces in favor of gas furnaces not only caused short-term pain but also lessened the long-term gains. What is missing here is an evidence-based, synergistic, multisectoral approach to pollution control. Combustion of natural gas produces NO_x, which not only is the major contributor to PM2.5 pollution but also acts as a precursor of tropospheric ozone in the atmosphere. Low-NO_x burners or NO_x emission control measures are yet to be used in the existing gas-fired boilers in China.[148] Increased NO_x concentrations

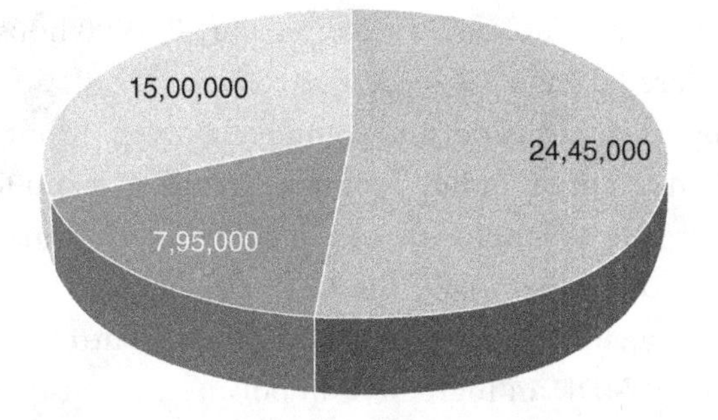

Figure 5.7 Number of households that have completed the conversion from coal to cleaner fuel in the BTH region.
Source: "Huanbaobu zoufang wubai yu wan hu" (MEP interviewed more than five million households), Zhongqing zaixian, December 24, 2017. www.chinanews.com/gn/2017/12–24/8408014.shtml.

produced by the coal-to-gas conversion program, combined with NO_x emitted by vehicles that run on fossil fuels, pose challenges to fundamentally addressing smog issue in Beijing.[149]

Moreover, the rush to substitute natural gas for coal pushed China's demand for gas beyond its capacity. A RAND study had predicted that China would need to acquire an additional 88 billion cubic meters of natural gas.[150] But the 2017 campaign resulted in a surge well beyond the estimates. According to NDRC, total gas consumption during January–August rose 17.8 percent from a year earlier, outstripping a 10.8 percent production increase.[151] The rising demand combined with unexpectedly low imports left China with a significant shortfall in the winter.[152] Not surprisingly, Hebei, with a gap between supply and demand as high as 20 percent, was the province hardest hit by the winter gas shortage. After the gas company began to ration gas, Hebei University-affiliated hospital warned that the restricted gas supply risked delay for surgeries and cross-infection in the hospital, even infectious disease outbreaks.[153] With coal banned and gas unavailable, many households in the northern provinces were left without heat in the winter.[154] In Zhouzhou village in Hebei province, 300 households were found to have no heating for several weeks in temperatures as low as –6°C (21°F).[155]An MEP survey found that of the 1,208 villages that completed the conversion, 426,000 households (5.6 percent) suffered from a gas shortage problem.[156]

Meanwhile, gas supplies to industrial and commercial users throughout China were curtailed or cut off in hopes of meeting the demand for civilian use. The shortage of gas supplies forced some industrial firms to suspend production. In December 2017, the German chemical giant, BASF, invoked "force majeure" (i.e., forces beyond its control) for a disruption in supplies of MDI, an ingredient in polyurethane used in packaging, from its facility in Chongqing.[157] To increase gas imports, government oil companies have been investing billions of dollars in building pipelines from gas fields in Russia and Central Asia. Sensing China's soaring demand for gas, gas suppliers in Central Asia (especially Turkmenistan and Uzbekistan) withheld the supply in hopes of getting better deals in other markets, which led them to break the terms of their contracts with China. By the end of January 2018, the volume of China's liquefied natural gas (LNG) through Central Asian pipeline networks had fallen by nearly half, from 150 million cubic meters to 70 million cubic meters. The sudden cuts of LNG by central Asian countries posed additional risk to China's energy security, forcing it to re-evaluate its energy supply strategy in Eurasia.[158]

Amid growing public outcry over the coal-to-gas conversion program, MEP issued a "double-urgent" letter to the twenty-eight northern cities on December 4, 2017. Saying that "keeping people warm in winter should be the No. 1 principle," the letter allowed households that had not completed the conversion to resume coal-fired heating.[159] The MEP letter was followed by a circular issued by NDRC and NEA three weeks later, which promoted the use of "clean coal," which is supposed to have a low sulfur content, to relax the gas shortage problem.[160] Despite the U-turns on coal ban, some localities continue to push for the conversion. In November 2018, Yingze District of Taiyuan, the capital of Shanxi province, was found to have imposed a blanket ban on burning coal meant to force local residents to adopt electric-powered heating. Without access to coal but unable to afford the new heating method, local residents tried to keep themselves warm by burning old furniture, discarded wood floors, and deadwood, sending up clouds of smoke that only worsened pollution in the city.[161] In December 2019, the government's push to burn clean coal also suffered a setback after at least six people who used the fuel for heating died in the city of Tangshan in Hebei province.[162]

SHORTCOMINGS OF AUTHORITARIAN DECISION-MAKING. The mixed outcome in addressing air pollution highlights the constraints and flaws of China's environmental governance model. According to a Chinese scholar at the Central Party School in Beijing, "the party's authoritarian leadership, the government's authoritarian regulation, and central authoritarian governance" have undermined the effectiveness of China's pollution control. To be specific:

> [They] not only constrain the power and ability of the subjects of three powers (the legislative body, the law enforcement agencies, and judiciary authorities), free market and local governments, but also result in the deviation from each other of the actions of environmental actors. The

> "leadership and implementation, differentiation and coordination, dominating and supporting, and execution and cooperation" relationship between the party and the state, inside the government, between the government and market, and between the government and society is in a state of reverse curvature, making it impossible to achieve a synergistic relationship between the aforementioned subjects and actors.[163]

The mixed outcome also reveals a deep dilemma in China's public policy process. Under Mao, ego-driven policymaking was responsible for many mistakes and disasters. Drawing lessons from this era, post-Mao leaders have made great efforts to reform the policymaking and implementation processes. As early as 1980, Deng Xiaoping called for reforming the party-state's decision-making system. His top associate, Vice Premier Wan Li, specified in 1986 that the fundamental goal of the reform was to achieve democratic and scientific decision-making.[164] As shown in China's success in reducing PM2.5 concentration, the increasing transparency and responsiveness have indeed improved policy outcomes. Still, it overstates matters to say that China is moving toward a more flexible and effective environmental governance regime. On the contrary, decades of reform and opening up have not fundamentally changed the impromptu, non-participatory, unaccountable, and mobilizational policy process, which often leads to undesirable and unintended policy outcomes. As two Chinese scholars observed, the complex issues ignored in the decision stage for rapid policy response can return at the implementation stage to compromise the final policy outcome in China.[165]

The process is vividly captured by the Chinese saying: "[Policymakers] make decisions by patting the head, make assurances by patting the chest, regret the decision by patting the legs, and walk away by patting the butt" (*pai naodai juece, pai xiongpu baozheng, pai datui houhui, pai pigu zouren*).

Conclusion

IT WAS NOVEMBER 14, 2018, ONE DAY AFTER MY ARRIVAL IN BEIJING. I was greeted with the worst air quality in a year-and-a-half: according to US Embassy data, the PM2.5 concentration soared to 288 by midday, higher than the levels recorded in several California cities that were suffering raging wildfires.[1] Hu Xijin, the editor in chief of China's nationalist tabloid *Global Times*, commented on social media:

> The smog today is so heavy. The [air quality] this year and last year was not bad. [I thought] the smog problem had been fixed. Yet the omnipresent smog is a big setback . . . Hope we will not have much smog in the coming winter, otherwise I will be speechless.

Two days later, a government newspaper published an article proclaiming that Beijing's average PM2.5 concentration dropped to a low of 49 $\mu g/m^3$ for the first ten months of the year, not seen since it began tracking the data. It also reported dramatic declines in SO_2, NO_2, and PM10 levels, as well as steady improvement in water quality and soil environment.[2]

Neither report was necessarily wrong (Beijing's smog is always worse in winter), but the repeated pollution crises point to the challenges China will continue to face as its authoritarian government attempts to overcome the environmental disaster produced by four decades of relentless industrial expansion and economic growth. Air, water, and soil pollution not only burden China with a colossal public health challenge, but they also take a heavy toll on China's society, economy, and polity. The health and non-health consequences of pollution, as well as the complexities involved in addressing the problem, will test the resilience of the Chinese state and its dreams of global leadership.

Interestingly, while policy elites in Beijing and Washington debate whether China's rise threatens international peace and security, they all seem to take China's rise as a given, overlooking the vast internal challenges it faces. Environmental health represents just one significant obstacle, though it may be the most formidable. In James Fallows' words, "environmental sustainability in all forms is China's biggest emergency, in every sense: for its people, for its government, for its effect on the world."[3] One thing that is very much lacking in the current debate in the foreign policy community is a clear understanding of how and why the environmental health crisis so seriously hinders China's rise. First, the people in China are more adversely affected by environmental degradation than those of almost any other country (India being its chief rival for that ignominious prize). Pollution's harmful effect on public health is still unfolding, but already hundreds of thousands of people die from pollution-related illnesses every year and tens of millions more suffer from pollution's chronic, debilitating effects. Simply put, China cannot regain its greatness in the world if its people continue to breathe polluted air, drink toxic water, and eat tainted food.

Second, China's economy, for all its rapid growth, has been harmed by pollution. Air pollution alone might cost China as much as 18 percent of its GDP in 2013. The price tag might be even larger when we consider the economic costs of water and soil pollution. Pollution also reduces grain yield, requiring greater and greater imports of food. Meanwhile, pollution-related health concerns hamper China's efforts to recruit and retain talents that are essential for sustainable economic growth and global financial hub building. The ensuing economic slowdown increases the likelihood China will slip into the middle-income trap, which occurs when a country makes significant gains in reducing poverty and building prosperity but then falters in achieving a fully developed status equal to advanced industrial economies.

Third, environmental health problems threaten sociopolitical stability by fueling widespread dissatisfaction and frustration among the populace, a breeding ground for mass protests directed at the state. In the post-Mao era, the Chinese Communist Party (CCP) needs to continuously justify its rule by not just sustaining rapid growth but also consistently accomplishing concrete policy goals. As their prosperity increases, the Chinese people increasingly value good health and other things beyond basic earnings. As a result, the gap between popular expectations and the state's capacity to deliver promised gains only becomes wider. Under Xi Jinping, the party-state dominance has reached its zenith in the post-Mao era, so government failure to deliver clean air, safe water, and untainted food may well lead people to question not only the rule of individual leaders but also the legitimacy of the political system itself. The lack of accountability and inability to change polluters' behavior increases the likelihood of petitions and protests escalating into large-scale violence, potentially raising a challenge to the legitimacy of the CCP.

Fourth, the crisis raises deep doubts about China's claims to assume the mantle of global leadership. Mounting environmental health challenges testify to the dark side of China's economic growth and undermine China's ability to market its development model internationally. The spillover effects of China's environmental health problems and concerns about whether China's vaunted Belt and Road Initiative is a backdoor way to outsource pollution also suggest it is not yet prepared to be a responsible leader in the international system. Furthermore, the crisis joins other internal challenges to tie the hands of Chinese foreign policymakers in mobilizing domestic support and resources in shouldering international responsibilities and projecting its power abroad. In 2014, a popular Chinese social media post joked that China was already a superpower in the following areas: pollution, corruption, toxic food, dropout students, stability-maintenance expenses, number of government officials, number of petitioners, and emigration.[4] Leading international relations scholars, including Richard Haass and John Mearsheimer, contend that it is still too early to view China as a global superpower or hegemon due to China's tremendous domestic challenges.[5] Their assessment of China's power capabilities is substantiated by Jessica Chen Weiss's research that domestic political imperatives continue to define China's foreign policy behavior, and China's efforts to seek global leadership reflect less a grand strategic ambition to overturn the liberal international order than the CCP leadership's desire to preserve its domestic rule.[6]

Finally, despite a wholesale commitment to the cause, the government thus far has achieved only mixed success in bringing pollution under control. Under President Xi, the state has enacted new laws, policies, regulations, and action plans to tackle environmental degradation. It also has introduced a number of new policy instruments seeking to align bureaucratic incentives with central policy goals in the implementation stage. The state commitment, however, is undermined by the gaps and deficiencies in environmental health-related policymaking as well as the perverse incentives it has created in a headlong rush to report gains on the pollution front. While the government has started to turn the tables on air pollution, the progress is uneven, the health benefits of pollution control remain limited, and the single-minded pursuit of pollution control has incurred unintended and undesirable outcomes on other policy fronts. The limited success of China's pollution control therefore highlights the problems arising from environmental authoritarianism, undercuts its ability to be a global leader in pollution control, and, more broadly, pinpoints the weaknesses of China's political leadership.[7]

THE STATE OF THE CHINESE STATE

An overview of China's evolving response to environmental health challenges suggests that government leaders are learning and responding, even if their learning curve is steep and they are not yet formally accountable to the people when making environmental and health-related policies. Policy development in China is largely crisis-driven: PM2.5 pollution did not seriously attract the attention of government leaders until it was recognized as an immediate and present danger that threatened socioeconomic stability as well as people's health. The elevation of environmental health onto the top leaders' agenda was facilitated by a socialization process that exposes Chinese policymakers to new concepts, discourse, policy instruments, and best practices. As a result, the government not only has upped the ante in pollution control but also has made assiduous efforts to improve policy design and implementation. This sets China apart from India, which tops the world in bad air quality. Despite having nine of the ten most polluted cities in the world, India did not adopt a national strategy to tackle air pollution until January 2019.[8]

In recognition of the tremendous challenges of translating policy statements and plans into actionable items, the Chinese government has focused on improving the accountability of government officials by making policy targets more quantifiable and transparent (e.g., construction of a nationwide air quality monitoring network). It also has introduced an array of policy instruments and tools to increase the cost of inaction and/or deviation from central preferences (e.g., the widespread use of inspection tours and disciplinary measures). In so doing, the central state has demonstrated relatively strong autonomy in decision-making *and* the ability to transmit pressures down to the lowest level of the hierarchy to get things done.

Despite a massive and complex PM2.5 crisis, the state has managed to mobilize the bureaucracy and make clear progress in bringing down the overall pollution level in a relatively short period. Furthermore, through the implementation of the new environmental protection law and three anti-pollution central action plans, the government has put in place distinctive environmental governance mechanisms, including centralized environmental monitoring, policy enforcement supervision, and river chief systems. To the extent that pro-environment policies in Western countries have had a dramatic impact on pollution since the 1970s, it is fair to assume that these state-driven measures, if pursued in a sustainable and effective manner, should ensure the improvement of China's environmental health in the long run. The implication is clear: the Chinese state has not exhausted its potential and any prediction about the imminent demise of the CCP regime would be a misread of the true nature of the state.

Many pundits and officials go further, however, contending that the country's experience in pollution control not only attests to the resilience of the Chinese state, it points to the superiority of the China model. Reasoning along this line, the former Minister of Environmental Protection Chen Jining anticipated that it would take less time for China to solve its air pollution problem than developed countries, which had spent two to four decades on overcoming the worst ravages of their environmental health problems.[9] Some opinion leaders in the West agree. "In the past four years, China has succeeded in cutting concentrations of one pollutant – fine particulates – by 32 percent, roughly what it took the United States 12 years to achieve after passage of the Clean Air Act in 1970," a Wall Street executive wrote in a *New York Times* article, indicating "China's version of capitalism" is winning.[10]

But such claims miss the fact that China's current resilience is derived not so much from institutional innovations or adaptations as from the strengthening of a mobilizational state.[11] Under Xi, succession politics has become more personalized, selection of political elites is based more on political loyalty than on professional competence, political participation has been neither institutionalized nor enlarged, and functional specialization increasingly is giving way to the all-around control by the party. In contrast to Maoist China, though, the opaque and seemingly exclusive authoritarian structure has not prevented outside actors such as foreign countries and international agencies from influencing the public policy discourse, which helps make the government more responsive to public demands while facilitating the adoption of internationally recognized norms, practices, and technologies. Also, unlike the ad hoc mass campaigns under Mao, Xi's China has invested significantly in institutionalizing and routinizing policy instruments introduced at the campaigns to tighten the grip of the party-state, a pattern some China scholars describe as "institutionalized mobilization."[12]

Does that make the China model a viable alternative to liberal democracy? No. Contrary to what Daniel Bell and others have argued, the weaknesses of the Chinese state are as glaring as its strengths.

First, the legitimacy problem continues to haunt the state due to its inability to anchor its rule in solid public approval that places less emphasis on performance and more on rule of law. Unbridled economic growth has imposed significant economic and social costs. Under its implicit deal derived from performance-based legitimacy, government failure to address these costs to the satisfaction of the populace can trigger a legitimacy crisis for the CCP regime. The hidden fragility of the state explains why Xi has attached such high importance to the environmental issue.[13]

Second, the lack of enlarged, institutionalized public participation in the policy process widens the gap between policy measures and people's actual wants and needs. Not surprisingly, despite government commitment to pollution control, Chinese officials have yet to unequivocally define access to a clean environment or the right to health as a fundamental human right, nor have they framed pollution as an explicit health problem.

Third, under the China model, the implementation of even a top policy issue such as pollution control remains lopsided, non-synergistic, and based on coercion. Re-creation of the bandwagon polity under Xi Jinping has forced government officials to take speedy action, but the unrelenting reliance on campaign-style implementation has distorted the bureaucratic incentive structure, leading to unwanted and unintended reverberations.

Finally, the health and non-health consequences of China's environmental destruction serve as a reminder that, despite China's official statements on the superiority of its development model, it is subject to the same negative consequences of countries that industrialized before it did and is paying an even greater price for its pollution. Such a model, if exported to other countries, is not likely to be popularized as best practice for governing environmental health or other development issues.

The deficiencies and difficulties authoritarian China faces in addressing its environmental health crisis support Thomas Jefferson's thesis that sick political systems are ultimately responsible for sick populations.[14] Indeed, a recent study conducted by Thomas Bollyky and collaborators used panel data covering 170 countries during a 46-year span (1970–2016) to demonstrate that having a liberal democracy matters more than the size and scale of the economy in addressing public health challenges, especially non-communicable diseases and injuries.[15]

Given the flawed implementation of environmental policy, can China accomplish its stated goals without an ultimate transition to democratic governance?[16] It's doubtful. Against the wish of reform-minded Chinese leaders in the 1980s, China has not fundamentally changed the non-scientific and nondemocratic aspects of its public policy process. To borrow the words of Michel Foucault, the government approach to pollution control is not characterized by "governmentality." Rather than promote the willing participation of the governed and exercising political power through different agencies, social groups, and techniques, the Chinese state still heavily relies on the formal bureaucracy and the exercise of sovereign power in the policy formulation and implementation processes.[17] Most of the "innovations" the CCP has introduced in the anti-pollution campaigns seek to improve top-down accountability and therefore are no more than efforts to tighten authoritarian control.

This lack of fundamental change points to the "systematic underdevelopment of institutions of governance among state and society at large," well documented by Carl Minzer in his book *End of An Era*.[18] It is fair to say that China has arrived at what Benjamin Liebman and Curtis Milhaupt considered "the worst of both worlds": the environmental health crisis suggests China no longer is insulated from the problems facing industrialized countries and that the Chinese state ultimately lacks the innovative institutional capacity to confront such challenges.[19]

In short, China's environmental health problems reveal a political system that is remarkably resilient but fundamentally flawed. The "perfect dictatorship" is not so perfect. As Stein Ringen concluded: "The present Chinese regime . . . is less strong, more dictatorial, and more of its own kind than the world has mostly wanted to believe."[20] A political system that is overwhelmingly tailored to Xi Jinping is not only a violation of the criteria set by Xi himself in terms of democracy and effective governance.[21] As shown in the COVID-19 outbreak, it is also highly susceptible to disruption and shocks.[22]

This is not to predict an imminent collapse of the Chinese state. Beijing may still muddle through the environmental crisis – as seen in its war on pollution and COVID-19, the government retains its ability to respond to a crisis by mobilizing resources and bureaucratic capacities for high-priority actions and has not exhausted its potential to make its institutional settings more conducive to policy implementation. But government leaders, preoccupied by pollution, health, and other mounting domestic challenges, will have to fight an uphill battle to maintain their grip on power and fulfill their global ambitions. As China's economic slowdown deepens, so will the tension between sustaining economic growth and environmental protection. Efforts to avoid the middle-income trap threatens to wear away the existing gains of pollution control while discouraging efforts to bring down the pollution levels further. Unfortunately, the archaic governance structure in China does not help ease this dilemma. The environmental crisis thus will continue to be one of the biggest obstacles to China's future economic growth and political stability. The inability to tackle such internal challenges may prompt the state to act more aggressively overseas to whip up nationalist sentiment and deflect domestic criticism in an attempt to shore up its political legitimacy. What we are likely going to witness is a faltering China that is perhaps more dangerous than a rising one.[23] In that sense, the environmental crisis could be the Achilles heel of modern China.

Notes

INTRODUCTION

1. Liu Zhen, “Endangered Bird Faints from Hunger after Failing to Catch Prey in China’s Choking Smog,” *South China Morning Post*, January 6, 2017. www.scmp.com/news/china/society/article/2059895/endangered-bird-faints-hunger-after-failing-catch-prey-chinas.
2. Katie Hunt, “Canadian Start-up Sells Bottled Air to China, Says Sales Booming,” *CNN*, December 15, 2015. www.cnn.com/2015/12/15/asia/china-canadian-company-selling-clean-air/index.html.
3. Ministry of Environmental Protection, *2015 nian zhongguo huanjing zhuangkuang gongbao* (The State of China’s Environment, 2015), May 2016. www.gov.cn/xinwen/2016–06/02/5078966/files/9ab14b4ce3294d5ab212bc83d3d31b7b.pdf.
4. Jin Yu, “Huanbaobu fabu yanjiu jieguo: woguo 2.8 yi jumin shiyong bu anquan yinyong shui” (“Ministry of Environmental Protection: 280 Million Residents in Our Country Do Not Have Access to Safe Drinking Water”), *Xinjing bao (Beijing News)*, March 15, 2014. http://politics.people.com.cn/n/2014/0315/c1001-24642042.html.
5. See Elizabeth C. Economy, *The River Runs Black: The Environmental Challenge to China’s Future* (Ithaca, NY: Cornell University Press, 2004).
6. Phil Brown, *Toxic Exposures: Contested Illnesses and the Environmental Health* (New York, NY: Columbia University Press, 2007), 1.
7. Literature addressing environmental health politics includes Peter Brimblecombe, *The Big Smoke: A History of Air Pollution in London since Medieval Times* (New York, NY: Methuen & Co., 1987); Timothy S. George, *Minamata: Pollution and the Struggle for Democracy in Postwar Japan* (Cambridge, MA: Harvard University Press, 2001); Peter Thorsheim, *Inventing Pollution: Coal, Smoke, and Culture in Britain since 1800* (Athens, OH: Ohio University Press, 2006); Chip Jacobs, *Smogtown: The Lung-Burning History of Pollution in Los Angeles* (New York, NY: Overlook Press, 2008); Brett L. Walker, *Toxic Archipelago: A History of Industrial Disease in Japan* (Seattle, WA: University of Washington Press, 2010).
8. Dr. Bates Gill, Remarks, Washington, DC, September 25, 2002.
9. See, for example, Economy, *The River Runs Black.*; Robert Stowe England, *Aging China: The Demographic Challenge to China’s Economic Prospects* (Washington, DC: Praeger, 2005); Yasheng Huang, *Capitalism with Chinese Characteristics: Entrepreneurship and the State* (New York, NY: Cambridge University Press, 2008); Carl E. Walter and Fraser J. T. Howie, *Red Capitalism: The Fragile Financial Foundation of China’s Extraordinary Rise* (Singapore: John

Willey & Sons, 2011); Bates Gill, *China's HIV/AIDs Crisis: Implications for Human Rights, the Rule of Law and U.S.-China Relations* (Washington, DC: CSIS, 2012); Judith Shapiro, *China's Environmental Challenges* (London: Polity Press, 2016); Minxin Pei, *China's Trapped Transition: The Limits of Developmental Autocracy* (Cambridge, MA: Harvard University Press, 2006); Minxin Pei, *China's Crony Capitalism: The Dynamics of Regime Decay* (Cambridge, MA: Harvard University Press, 2016).

10. See Judith Shapiro, *Mao's War against Nature: Politics and the Environment in Revolutionary China* (New York, NY: Cambridge University Press, 2001).
11. Karl Polanyi, *The Great Transformation: The Political and Economic Origin of Our Time* (Boston, MA: Beacon Press, 1944); Economy, *The River Runs Black;* Yanzhong Huang, *Governing Health in Contemporary China* (New York, NY: Routledge, 2013).
12. Yanzhong Huang, "China: The Dark Side of Growth," *Yale GlobalOnline*, June 6, 2013. https://yaleglobal.yale.edu/content/china-dark-side-growth.
13. Henry Kissinger, *On China* (New York, NY: Penguin, 2011), 11–12.
14. Ivo H. Daalder and James M. Lindsay, *The Empty Throne: America's Abdication of Global Leadership* (New York, NY: PublicAffairs, 2018).
15. Catherine Wong, "Xi Jinping Portrays China as Global Leader as Donald Trump Prepares to Take Office," *South China Morning Post*, January 19, 2017. www.scmp.com/news/china/diplomacy-defence/article/2063584/xi-portrays-china-global-leader-trump-prepares-take.
16. Ministry of Foreign Affairs of the People's Republic of China, "Xi Jinping Meets with IMF Managing Director Christine Lagarde," April 10, 2018. www.fmprc.gov.cn/mfa_eng/zxxx_662805/t1550191.shtml.
17. "Shangjiang: nongcun wuran yi weixie dao zhengbing" ("General: Pollution in the Countryside Has Threatened Conscription"), *Xinjin bao (Beijing News)*, March 13, 2016. https://xw.qq.com/mil/20160313014764/MIL2016031301476403.
18. Murry Feshbach and Alfred Friendly, Jr., *Ecocide in the USSR: Health and Nature under Siege* (New York, NY: BasicBooks, 1992).
19. "Will China Crumble?" *Foreign Affairs*, April 30, 2015. www.foreignaffairs.com/articles/china/2015-04-30/will-china-crumble.
20. Jared Diamond, *Collapse: How Societies Choose to Fall or Succeed* (New York, NY: Penguin, 2005).
21. The Lex Column, "Belt and Road a Way for China to Export Bad Air," *Financial Review*, May 29, 2017. www.afr.com/news/politics/world/belt-and-road-a-way-for-china-to-export-bad-air-20170529-gwf8ea; Jean Chemnick, "To Cut Emissions, China's Global Infrastructure Plan May Need a Greener Path," *Scientific American*, April 26, 2019. www.scientificamerican.com/article/to-cut-emissions-chinas-global-infrastructure-plan-may-need-a-greener-path/.
22. Elizabeth Economy, "The Great Leap Backward," *Foreign Affairs*, September/October 2007.
23. Jennifer Holdaway, "Environment and Health Research in China: The State of the Field," *The China Quarterly* 214 (June, 2013): 256.
24. Jennifer Holdaway and Wang Wuyi, "Stronger Enforcement Won't Be Enough to Solve China's Environment and Health Problems," *chinadialogue*, April 29, 2014. www.china

dialogue.net/article/show/single/en/6926-Stronger-enforcement-won-t-be-enough-to-solve-China-s-environment-and-health-problems.

25. Thomas Johnson, "The Health Factor in Anti-Waste Incinerator Campaigns in Beijing and Guangzhou," *The China Quarterly* 214 (June, 2013): 363.
26. Holdaway and Wang, "Stronger Enforcement Won't Be Enough to Solve China's Environment and Health Problems."
27. Chen Wei, "Interpreting PM2.5 'Gene Spectrum': Major Differences Exist between Beijing and Shanghai on Pollution Source," *Jiefang ribao (Liberation Daily)*, January 18, 2015. www.chinanews.com/gn/2015/01–18/6978832.shtml.
28. Yu Yunjiang, interview with author, South China Institute of Environmental Sciences, Guangzhou, November 23, 2015.
29. Officials at Jiangsu provincial CDC, Nanjing, interview with author, December 18, 2015.
30. Holdaway, "Environment and Health Research in China: The State of the Field," p. 258.
31. National Joint Center for Air Pollution Prevention and Control, "Zhuanjia jiedu" ("Expert Interpretation"), March 4, 2019. https://finance.sina.com.cn/roll/2019–03-04/doc-ihsxncvf9589181.shtml.
32. Bryan Tilt, *Dams and Development in China: The Moral Economy of Water and Power* (New York, NY: Columbia University Press, 2015), 81. See also Yan Yunxiang, *The Individualization of Chinese Society* (London: London School of Economics Monographs on Social Anthropology, 2019).
33. Anna Lora-Wainwright, "Dying for Development: Pollution, Illness and the Limits of Citizens' Agency in China," *The China Quarterly* 214 (June, 2013): 244; Sam Geall (ed.), *China and the Environment: The Green Revolution* (London; New York: Zed Books, 2013) documented a variety of strategies and tactics used by journalists, lawyers and ordinary Chinese in their struggle against health-harming pollution.
34. Yang Guobin, "Contesting Food Safety in Chinese Media: Between Hegemony and Counterhegemony," *The China Quarterly* 214 (June, 2013).
35. Kevin O'Brien and Lianjiang Li, *Rightful Resistance in Rural China* (Cambridge, MA: Cambridge University Press, 2006).
36. Shi Tianjian, *Political Participation in Beijing* (Cambridge, MA: Harvard University Press, 1997).
37. Bryan Tilt, *The Struggle for Sustainability in Rural China: Environmental Values and Civil Society* (New York, NY: Columbia University Press, 2010); Deng Yanhua and Guobin Yang, "Pollution and Protest: Environmental Mobilization in Context," *The China Quarterly* 214 (June, 2013); Anna Lora-Wainwright, "The Inadequate Life: Rural Industrial Pollution and Lay Epidemiology in China," *The China Quarterly* 214 (May, 2013).
38. Richard Wike and Bruce Stokes, "Chinese Public Sees More Powerful Role in World, Names U.S. as Top Threat," Pew Research Center, October 5, 2016. www.pewglobal.org/2016/10/05/chinese-public-sees-more-powerful-role-in-world-names-u-s-as-top-threat/.
39. Michael Mann, *The Sources of Social Power, Volume II: The Rise of Classes and Nation-States, 1760–1914* (New York, NY: Cambridge University Press, 1993), 59.

40. Theda Skocpol, "Bringing the State Back In: Current Research," in *Bringing the State Back In*, edited by Peter B. Evans, Dietrich Rueschemeyer, and Theda Skocpol (New York, NY: Cambridge University Press, 1985).
41. Dali L. Yang, *Remaking the Chinese Leviathan* (Stanford, CA: Stanford University Press, 2004); Yasheng Huang, *Investment and Inflation Controls in China* (New York, NY: Cambridge University Press, 1996).
42. Jean Oi, "Fiscal Reform and the Economic Foundations of Local State Corporatism in China," *World Politics* 45 (October, 1992): 99–126; Andrew Walder, "Market and Inequality in Transitional Economies: Toward Testable Theories," *American Journal of Sociology* 101, no. 4 (January, 1996): 1060–1073.
43. Lora-Wainwright, "The Inadequate Life: Rural Industrial Pollution and Lay Epidemiology in China," 302–320.
44. Bryan Tilt, "Industrial Pollution and Environmental Health in Rural China: Risk, Uncertainty and Individualization," *The China Quarterly* 214 (June, 2013): 283–301.
45. Jakob Klein, "Everyday Approaches to Food Safety in Kunming," *The China Quarterly* 214 (June, 2013): 376–393.
46. Lora-Wainwright, "Dying for Development: Pollution, Illness and the Limits of Citizens' Agency in China," 244.
47. Economy, *The River Runs Black*.
48. Feng Zhongcao (ed.), *Zhongguo huan jing yu jian kang bao gao* (*An Assessment of the Toxicity and Hazards of Pollutants in China*) (Beijing: Zhongguo huan jing ke xue chu ban she, 1999).
49. WHO and UN Development Programme, *Environment and People's Health in China*, 2001. www.wpro.who.int/environmental_health/documents/docs/CHNEnvironmentalHealth.pdf; World Bank and SEPA, *The Cost of Pollution in China: Economic Estimates of Physical Damage*, WHO (World Health Organization), 2005. CCICED (China Council for International Collaboration on Environment and Development), "Environment and Health Management System and Policy Framework," November 12, 2008, www.iisd.org/sites/default/files/publications/CCICED/responsibilities/2008/environment-health-management-policy-framework.pdf; Charles W. Freeman and Xiaoqing Lu, *Assessing Chinese Government Response to the Challenge of Environment and Health*, A report of the CSIS Freeman Chair in China Studies, Center for Strategic and International Studies (June, 2008). Since 2005, Friends of Nature has edited and produced annual reports on environment development of China. Titled, *Green Book of Environment* in China, the reports rely on facts and empirical data to highlight, among others, the environmental health challenges in China. English versions of the reports (*The China Environment Yearbook*) have been published by Brill since 2007.
50. Yanzhong Huang, "International Institutions and China's Health Policy," *Journal of Health Politics, Policy and Law* 40, no. 1 (February, 2015): 41–71. See also Susan Shirk, "Internationalization and China's Economic Reforms," in *Internationalization and Domestic Politics*, edited by Robert O. Keohane and Helen V. Milner, 186–206 (New York: Cambridge University Press, 1996); Fang Songying and Randall W. Stone "International Organizations as Policy Advisors," *International Organization* 66 (4) (2012): 537–569.

51. See David Lampton (ed.), *Policy Implementation in Post-Mao China* (Berkeley, CA: University of California, 1987).
52. Adam Przeworski, Susan C. Stokes, and Bernard Manin (eds.), *Democracy, Accountability, and Representation* (New York, NY: Cambridge University Press, 1999), 133–134.
53. See David Lampton, *The Politics of Medicine in China: The Policy Process, 1949–1977* (New York, NY: Westview Press, 1977).
54. Avery Goldstein, *From Bandwagon to Balance of Power: Structural Constraints and Politics in China 1941–1971* (Palo Alto, CA: Stanford University Press, 1991).
55. Zheng Yongnian, *De Facto Federalism in China: Reforms and Dynamics of Central-Local Relations* (Singapore: World Scientific, 2006), 38–39.
56. Barry Naughton, "The Decline of Central Control over Investment in Post-Mao China," in *Policy Implementation in Post-Mao China*, edited by David M. Lampton, 51–79 (Berkeley and Los Angeles, CA: University of California Press, 1987).
57. Huang, *Governing Health in Contemporary China.*
58. Economy, *The River Runs Black,* 121.
59. Maria Edin, "State Capacity and Local Agent Control in China: CCP Cadre Management from a Township Perspective," *The China Quarterly* 173 (March, 2003): 39.
60. Eun Kyong Choi, "Informal Tax Competition among Local Governments in China since the 1994 Tax Reforms," *Issues & Studies* 45, no.2 (June, 2009): 159–183.
61. Genia Kostka, "China's Local Environmental Policies," in *Routledge Handbook of Environmental Policy in China,* edited by Eva Sternfeld, 31–47 (New York, NY: Routledge, 2017); Huang Han, "Zhibiao zhili jiqi kunjing" ("Performance Target Management and Its Dilemma"), *Haerbing gongye daxue xuebao* (*Journal of Harbin Institute of Technology*), Social Sciences Edition 18, no. 6 (December, 2016): 37–45.
62. Goh Sui Noi, "19th Party Congress: Anti-graft Drive Not Causing Inertia among Officials, Says Party Official," *The Straits Times,* October 19, 2017. www.straitstimes.com/asia/east-asia/19th-party-congress-not-true-anti-graft-drive-led-to-inaction-among-officials.
63. Diamond, *Collapse: How Societies Choose to Fall or Succeed,* 377.
64. Samuel Huntington, *The Third Wave: Democratization in the Late Twentieth Century* (Noman, OK: University of Oklahoma Press, 1991); Francis Fukuyama, "The End of History?" *The National Interest* (1989).
65. Joshua Kurlantzick, *Democracy in Retreat: The Revolt of the Middle Class and the Worldwide Decline of Representative Government* (New Haven, CT: Yale University Press, 2013).
66. Larry Diamond, "When Does Populism Become a Threat to Democracy?" For the FSI Conference on Global Populisms at Stanford University, November, 2017.
67. Sean Illing, "20 of America's Top Political Scientists Gathered to Discuss Our Democracy. They're Scared," *Vox,* October 13, 2017. www.vox.com/2017/10/13/16431502/america-democracy-decline-liberalism
68. Larry Diamond, *Ill Winds: Saving Democracy from Russian Rage, Chinese Ambition, and American Complacency* (New York: Penguin Press, 2019).
69. For example, see David Shambaugh, *China's Communist Party: Atrophy and Adaptation* (Washington, DC: Woodrow Wilson Center Press, 2008), 176.

70. Andrew Nathan, "China's Changing of the Guard: Authoritarian Resilience," *Journal of Democracy* 14, no. 1 (January, 2003): 6–7.
71. Kerry Brown, *Friends and Enemies: The Past, Present and Future of the Communist Party of China* (New York, NY: Anthem Press, 2009); Bruce Dickson, *Red Capitalists in China: The Party, Private Entrepreneurs, and Prospects for Political Change* (New York, NY: Cambridge University Press, 2003); Bruce Dickson, *Wealth into Power: The Communist Party's Embrace of China's Private Sector* (New York, NY: Cambridge University Press, 2008); Kellee Tsai, *Capitalism without Democracy: The Private Sector in Contemporary China* (Ithaca, NY: Cornell University Press, 2007); Shambaugh, *China's Communist Party*.
72. Bruce Gilley, "The Limits of Authoritarian Resilience," *Journal of Democracy* 14, no. 1 (January, 2003).
73. Andrew Nathan, "China Since Tiananmen: Authoritarian Impermanence," *Journal of Democracy* 20, no. 3 (July, 2009): 39.
74. See David Shambaugh, "The Coming Chinese Crackup," *The Wall Street Journal*, March 6, 2015; David Shambaugh, *China's Future?* (London: Polity, 2016); Pei, *China's Crony Capitalism*.
75. Cheng Li, "The End of the CCP's Resilient Authoritarianism? A Tripartite Assessment of Shifting Power in China," *The China Quarterly* 211 (September, 2012): 595–623.
76. Shambaugh, "The Coming Chinese Crackup."
77. Pei, *China's Trapped Transition*; Huang, *Capitalism with Chinese Characteristics*.
78. Pei, *China's Crony Capitalism*.
79. Joshua Kurlantzick, *State Capitalism: How the Return of Statism is Transforming the World* (New York, NY: Oxford University Press, 2016).
80. Mark Thompson, *Authoritarian Modernism in East Asia* (New York, NY: Palgrave Macmillan, 2019).
81. Eric Li, "The Life of the Party: The Post-Democratic Future Begins in China," *Foreign Affairs*, January/February 2012.
82. Daniel A. Bell, *The China Model: Political Meritocracy and the Limits of Democracy* (Princeton, NJ: Princeton University Press, 2016).
83. Nicholas R. Lardy, *The State Strikes Back: The End of Economic Reform in China?* (New York, NY: Columbia University Press, 2019); Carl Minzner, *End of An Era: How China's Authoritarian Revival is Undermining Its Rise* (New York, NY: Oxford University Press, 2018); Yu Yu, "CPC: Cadres Should Not Be Echoes of Western Moral Values," *BBC News*, July 20, 2014. www.bbc.com/zhongwen/simp/china/2014/07/140720_china_propaganda_common_values.shtml.
84. For an incisive overview of the transformative changes underway in Xi's China, see Elizabeth Economy, *The Third Revolution: Xi Jinping and the New Chinese State* (New York, NY: Oxford University Press, 2018).
85. Yu Keping, Remarks at the Third Conference on Peace and Development, New York, July 15, 2014.
86. Li Laifang, "Commentary: Enlightened Chinese democracy puts the West in the shade," *Xinhua*, October 17, 2017. www.xinhuanet.com/english/2017–10/17/c_136685546.htm.

87. "Socialism with Chinese Characteristics Enters New Era: Xi," *Xinhua*, October 18, 2017. http://news.xinhuanet.com/english/2017-10/18/c_136688475.htm.
88. Charlotte Gao, "For the First Time, Chinese Communist Party to Hold a World Political Parties Dialogue," *Diplomat*, November 29, 2017. https://thediplomat.com/2017/11/for-the-first-time-chinese-communist-party-to-hold-a-world-political-parties-dialogue/.
89. Yang, *Remaking the Chinese Leviathan*, 312.
90. Wen Jiabao, "Tongzhou gongji gongdu nanguan" ("Get through Tough Times Together"), in *Wen Jiabao tan jiaoyu* (*Wen Jiabao on Education*) (Beijing: Renmin chubanshe and Renmin jiaoyu chubanshe, 2013) http://health.sohu.com/20131101/n389393289.shtml.
91. See, for example, Kenneth Lieberthal and David Lampton (eds.), *Bureaucracy, Politics, and Decision Making in Post-Mao China* (California: University of California Press, 1992).

HEALTH EFFECTS OF ENVIRONMENTAL DEGRADATION

1. Fan Xinping, *Zhongguo jinghe jingji: jingji gaige zhenghe fenxi (China's Competing and Cooperative Economy: An Integrated Analysis of Economic Reform)* (Beijing: Zhonggong zhongyang dangxiao chubanshe, 1998).
2. Robert Wilson, "The Explosive Growth of Steel Production in China: Why It Matters,"*Carbon Counter,* June 9, 2015. https://carboncounter.wordpress.com/2015/06/09/the-explosive-growth-of-steel-production-in-china-why-it-matters/.
3. Andy Home, "China Has Created a Steel Monster and Now Must Tame It," *Reuters,* October 29, 2015. www.reuters.com/article/china-steel-ahome/column-china-has-created-a-steel-monster-and-now-must-tame-it-andy-home-idUSL8N12T3IV20151029.
4. Wang Erde, "Jingjinji daqi wuran zhili keyi chengwei quyu lvse fazhan de qiji" ("Air pollution control in Beijing-Tianjin-Hebei can be an opportunity for regional green development"), *Zhongguo huanjing guanli (China Environmental Management)* 9, no. 1 (2017): 18–20.
5. "Hebei manbao gangtie chanliang 5000 wandun" ("Hebei province underreported 50 million tons of steel"), *21 shiji jingji baodao (Financial Report of the 21st Century),* September 5, 2012. http://money.163.com/12/0905/02/8AJTET1E00253B0H.html.
6. The World Bank, "Energy Use (kg of Oil Equivalent per Capita)," 2014. https://data.worldbank.org/indicator/EG.USE.PCAP.KG.OE?end=2014&locations=CN&start=1971&view=chart.
7. Ana Swanson, "How China Used More Cement in 3 Years than the U.S. Did in the Entire 20th Century," *The Washington Post,* March 24, 2015. www.washingtonpost.com/news/wonk/wp/2015/03/24/how-china-used-more-cement-in-3-years-than-the-u-s-did-in-the-entire-20th-century/.
8. Sun Xiuyan, "Huanbaobu huiying wumai fangzhi" ("MEP's response on smog control"), *Renmin ribao* (*People's Daily*), January 28, 2013. http://cpc.people.com.cn/n/2013/0128/c83083-20345312.html.
9. Julian Zhu, Yan Yan, Christina He, et al., "China's Environment: Big Issues, Accelerating Effort, Ample Opportunities," *Goldman Sachs,* July 13, 2015, 24. www.goldmansachs.com/our-thinking/pages/interconnected-markets-folder/chinas-environment/report.pdf.

10. National Bureau of Statistics of China, *Zhonghua renming gongheguo 2016 nian guomin-jingji he shehui fazhan tongji gongbao* (*Statistical Communiqué of the People's Republic of China on the 2016 National Economic and Social Development*), February 28, 2017. www.stats.gov.cn/tjsj/zxfb/201702/t20170228_1467424.html.
11. "Woguo jidongche baoyouliang da 2.9 yi liang" ("Vehicle inventory in our country reaches 290 million"), *Xinhua News*, January 11, 2017. www.xinhuanet.com/politics/2017–01/10/c_129440197.htm.
12. Automotive Industry Portal Marklines, "China – Flash Report, Sales volume, 2017," December, 2017. www.marklines.com/en/statistics/flash_sales/salesfig_china_2017.
13. Wang Yuesi, "Yiwen kandong zhongguo wumai de chengyin weihai he jiejue fangan" ("The causes, harms and solutions regarding smog in China"), *Restoring Blue Skies*, 2013. www.efchina.org/News-zh/Program-Updates-zh/RestoringBlueSkies-zh/pop_science/pop_science_20160122.
14. Wang Erde, "Jingjinji daqi wuran zhili keyi chengwei quyu lvse fazhan de qiji."
15. Justin Worland, "9 out of 10 Chinese Cities Fail Pollution Test," *Time*, February 2, 2015. http://time.com/3692062/china-pollution-test/.
16. Ruixue Jia, "Pollution for Promotion," IIES, Stockholm University, October 21, 2012. www.dartmouth.edu/~neudc2012/docs/paper_58.pdf.
17. Factory owner, interview with author, January 11, 2019.
18. "Zhongguo xuezhe jiedu yihao wenjian he sannong wenti" ("Chinese scholar interprets document no. 1 and 'Three Rurual' issues"), *Ouzhou shibao (European times)*, August 4, 2017. www.oushinet.com/china/cnis/20170804/268606.html.
19. Shuping Niu, "China Needs to Cut Use of Chemical Fertilizers: Research," *Reuters*, January 14, 2010. www.reuters.com/article/us-china-agriculture-fertiliser/china-needs-to-cut-use-of-chemical-fertilizers-research-idUSTRE60D20T20100114.
20. Fu Yongjun, "Jingpao zai nongyao huafei li de guodu" ("A country soaked in pesticides and fertilizers"), *Guoke Luyuan*. http://bjgkly.com/newsinfo.asp?id=45.
21. Fu, "Jingpao zai nongyao huafei li de guodu."
22. Wang Ke, "Zhongguo mujun huafei yongliang shi meiguo 2.6 bei" ("China's average amount of fertilizer use per mu is 2.6 times of the U.S."), *Zhongguo jingji zhoukan* (*China Economic Weekly*) no. 34, 2017; "The Most Neglected Threat to Public Health in China Is Toxic Soil," *The Economist*, June 8, 2017. www.economist.com/news/briefing/21723128-and-fixing-it-will-be-hard-and-costly-most-neglected-threat-public-health-china.
23. Michael Standaert, "China to Ban 10 More Highly Toxic Pesticides," *Bloomberg*, December 6, 2017. www.bna.com/china-ban-10-n73014472829/.
24. Qiang Xiao, "Chinese Chemical Threat to Rivers," *BBC News*, January 24, 2006. http://news.bbc.co.uk/2/hi/asia-pacific/4642090.stm.
25. Jeffrey Hays, "Water Pollution in China," *Facts and Details*, April 2014. http://factsanddetails.com/china/cat10/sub66/item391.html.
26. Yang Jian, "China's River Pollution 'A Threat to People's Lives,'" *Shanghai Daily*, February 17, 2012. http://en.people.cn/90882/7732438.html.

27. Xu Dongqun, "Woguo huanjing yu jiankang mianlin de tiaozhan ji yingdui celue" ("China's Rivers: Frontlines for Chemical Wastes"), *Worldwatch Institute*, February 23, 2006. www.worldwatch.org/chinas-rivers-frontlines-chemical-wastes.
28. Xiao Yin Zhang, et al., "Atmospheric Aerosol Compositions in China: Spatial/Temporal Variability, Chemical Signature, Regional Haze Distribution and Comparisons with Global Aerosols," *Atmospheric Chemistry and Physics* 12, no. 2 (January, 2012): 779–799; Di Chang, et al., "Visibility Trends in Six Megacities in China 1973–2007," *Atmospheric Research* 94, no. 2 (October, 2009): 161–167. In the Mao era, most pollutants were concentrated in cities dominated by heavy industry. Professor Xu Jianhua, interview with author, July 2, 2015.
29. Tang Wentao, "Woguo daqi wuran yizhi hen Yanzhong" ("Air pollution in our country has always been serious"), March 25, 2015. http://blog.sina.com.cn/s/blog_5849bd280102vf6w.html.
30. "Air Pollution," *World Health Organization*. www.who.int/topics/air_pollution/en/.
31. Ministry of Environmental Protection of China, *2013 zhongguo huanjing zhuangkuang gongbao* (China Environmental Status Bulletin 2013), May 2014. www.mee.gov.cn/hjzl/zghjzkgb/lnzghjzkgb/201605/P020160526564151497131.pdf.
32. Pan Xiaochuan, Li Guoxing, and Gao Ting, *Weixian de huxi (Dangerous Breath: A Study of the Health Hazard and Economic Losses of PM2.5)* (Beijing: Huanjing kexue chubanshe, December 1, 2012). http://m.greenpeace.org/china/Global/china/publications/campaigns/climate-energy/2012/dangerous-breath.pdf.
33. The World Bank and the Government of the People's Republic of China, *Cost of Pollution in China: Economic Estimates of Physical Damages* (Washington, DC: World Bank, 2007). http://documents.worldbank.org/curated/en/782171468027560055/Cost-of-pollution-in-China-economic-estimates-of-physical-damages.
34. Robert A. Rohde and Richard A. Muller, "Air Pollution in China: Mapping of Concentrations and Sources," *PLOS ONE*, August 20, 2015. http://journals.plos.org/plosone/article?id=10.1371/journal.pone.0135749.
35. Robert Ferries, "China Air Pollution Far Worse Than Thought: Study," *CNBC*, August 18, 2015. www.cnbc.com/2015/08/18/china-air-pollution-far-worse-than-thought-study.html.
36. James Fallows, "2 Charts That Put the Chinese Pollution Crisis in Perspective," *The Atlantic*, April 18, 2014. www.theatlantic.com/international/archive/2014/04/2-charts-that-put-the-chinese-pollution-problem-in-perspective/360868/.
37. "What Is Soil Pollution?" *Environmental Pollution Center*. www.environmentalpollutioncenters.org/soil/.
38. "The Most Neglected Threat to Public Health in China is Toxic Soil," *The Economist*, June 8, 2017.
39. Ministry of Environmental Protection and Ministry of Land and Resources, *Quanguo turang wuran zhuangkuang diaocha gongbao* (Investigation report on national soil contamination situation), April 17, 2014. www.mee.gov.cn/gkml/hbb/qt/201404/W020140417558995804588.pdf.
40. Green Mining United Institute, "Boshike nongtian zhongjinshu gewuran zonghe fangzhi jishu yu duice" ("Boshike: Technologies and Countermeasures Regarding

Comprehensive Control of Heavy Metal Cadmium Pollution"), April 1, 2017. www.chinagmu.com/index.php/2017/04/01/case-5/.

41. Dr. Shang Qi, interview with author, China CDC, Beijing, July 1, 2015.
42. "The Most Neglected Threat to Public Health in China Is Toxic Soil," *The Economist*, June 8, 2017.
43. The State Council Information Office of the People's Republic of China, *Zhongguo jiankang shiye de fazhan yu renquan jinbu baipishu* (*The Whitepaper on the Development of China's Health Undertakings and the Progress of Human Rights)*, September 29, 2017. www.scio.gov.cn/zfbps/ndhf/36088/Document/1565111/1565111.htm?from=timeline&isappinstalled=0.
44. "Geren weisheng zhichu zhanbi jiangzhi 20 nianlai zuidi" ("The Share of Out of Pocket Payment Drops to the Lowest Level in 20 Years"), *Renmin ribao* (*People's Daily*), October 6, 2017. www.gov.cn/guowuyuan/2017-10/06/content_5229750.htm.
45. "Zhongguo manxingbinghuan wunian zeng jin yiyi ren" ("NCD Patients in China Increased 100 Million in the Past Five Years"), *Caixin*, June 8, 2016. http://m.china.caixin.com/m/2016-06-08/100952669.html.
46. World Health Organization, "China." WHO Noncommunicable Diseases Country Profiles (2018) www.who.int/nmh/countries/chn_en.pdf; World Health Organization, "United States." WHO Noncommunicable Diseases Country Profiles (2018). www.who.int/nmh/countries/usa_en.pdf.
47. Chen Jing, "Zhongguo xinxueguan jiankang zhishu chulu" ("Release of China Cardiovascular Health Index"), *China News Network*, May 26, 2017. www.chinanews.com/cj/2017/05-26/8235253.shtml.
48. Yanzhong Huang, "Older and Unhealthier," *China Economic Quarterly*, March, 2017.
49. Yin Shijie (ed.), "Guojia aizheng zhongxin: Feiai ju exing zhongliu fabing di yi wei" ("National Cancer Research Center: Malignant Lung Cancer Ranks no. 1 in Incidence Rate"), *Xinhuanet*, April 6, 2018. www.xinhuanet.com/politics/2018-04/06/c_1122643981.htm.
50. Between 1991 and 2015, cancer mortality rate in the United States dropped by 26 percent, resulting in nearly 2.4 million fewer deaths during that time. See American Cancer Society, "Cancer Mortality in the US Continues Decades-long Drop," January 4, 2018. www.sciencedaily.com/releases/2018/01/180104120258.htm.
51. Shan Juan, "Lung Cancer Rising, But Not from Smoking," *China Daily*, August 11, 2017. www.chinadaily.com.cn/china/2017-08/11/content_30451525.htm.
52. Li Qiumeng, "Siwang lu shangsheng baifenzhi 465" ("Mortality rate increased by 465 percent"), *Jinghua shibao* (*Beijing Times*), November 18, 2013. http://politics.people.com.cn/n/2013/1118/c70731-23574864.html.
53. Shan Juan, "Report Spells Out China's Cancer Risk," *China Daily*, January 1, 2013. www.chinadaily.com.cn/china/2013-01/10/content_16100330.htm; Guojia aizheng zhongxin (China National Cancer Center), "2018 nian quanguo zuixin aizheng baogao" ("The New National Cancer Report, 2018"), www.lascn.net/Item/75986.aspx.
54. Li Qiumeng, "Siwang lu shangsheng baifenzhi 465."

55. Wen Shuping, "2017 feiai fabingshu da 80 wan" ("Lung Cancer Incidences for 2017 Reached 800,000"), *Jingji guancha wang*, November 10, 2017. www.eeo.com.cn/2017/1110/316519.shtml.
56. Liwen Fang, Pei Gao, Heling Bao, et al., "Chronic Obstructive Pulmonary Disease in China: A Nationwide Prevalence Study," *The Lancet Respiratory Medicine* 6 (June, 2018).
57. Tu Jun, "Zhongguo siwang renshu jin jiuchen yin manxingbing" ("Nearly 90% of Mortality in China is Attributed to NCDs"), *The Paper*, March 26, 2017. www.thepaper.cn/newsDetail_forward_1647638; SL Murphy, J Xu, KD Kochanek, "Deaths: Final Data for 2010," *National Center for Health Statistics*, May, 2013. www.cdc.gov/nchs/data/dvs/deaths_2010_release.pdf.
58. Shu-Ching Jean Chen, "Mental Health Care: China's Next Big Market," *Forbes*, January 26, 2016. www.forbes.com/sites/shuchingjeanchen/2016/01/26/mental-health-care-chinas-next-big-market/#23c628f944fe.
59. The calculation was based on numbers provided by a 2017 China Public Health White Paper; Yun Gao, et al., "Prevalence of Hypertension in China: A Cross-Sectional Study," *PLOS ONE* 8, no. 6 (June, 2013). www.ncbi.nlm.nih.gov/pmc/articles/PMC3679057/.
60. Local health official, interview with author, Hebei province, November 14, 2018.
61. Yanzhong Huang, *Governing Health in Contemporary China* (New York: Routledge, 2013).
62. Anne-Emanuelle Birn, Yogan Pillay, and Timothy H. Holtz, *Textbook of International Health: Global Health in a Dynamic World* (New York, NY: Oxford, 2009), 285–287.
63. CSDH, "Closing the Gap in a Generation: Health Equity through Action on the Social Determinants of Health," Final Report of the Commission on Social Determinants of Health (Geneva: World Health Organization, 2008). http://apps.who.int/iris/bitstream/10665/43943/1/9789241563703_eng.pdf.
64. The World Health Organization, "An Estimated 12.6 Million Deaths Each Year are Attributable to Unhealthy Environments," Geneva: WHO, 2016. www.who.int/en/news-room/detail/15-03-2016-an-estimated-12-6-million-deaths-each-year-are-attributable-to-unhealthy-environments.
65. The World Health Organization, "An Estimated 12.6 Million Deaths Each Year are Attributable to Unhealthy Environments."
66. Katherine Allen and Corinne Burns, "Minamata Disease," Boston University Sustainability, December 14, 2009. www.bu.edu/sustainability/minamata-disease/.
67. See, for examples, C. Arden Pope III, et al., "Lung Cancer, Cardiopulmonary Mortality and Long-Term Exposure to Fine Particulate Air Pollution," *JAMA* 287, 9 (March, 2002): 1132–1141; Wei Huang, et al., "Air Pollution and Autonomic and Vascular Dysfunction in Patients with Cardiovascular Disease:Interactions of Systemic Inflammation, Overweight, and Gender," *Am J Epidemiol* 176, 2 (July, 2012): 117–126; Jack E. Fergusson, *The Heavy Elements: Chemistry, Environmental Impact, and Health Effects*, First Edition (Oxford, England: Pergamon Press, 1990); Industrial Economics, Inc., "Literature Review of Air Pollution -Related Health Endpoints and Concentration-Response Functions for Ozone, Nitrogen Dioxide, and Sulfur Dioxide: Results and Recommendations," Draft report for South Coast Air Quality Management District, October 2, 2015. www.aqmd.gov/docs/default-source/Agendas/STMPR-Advisory-Group/december-2015/3b_draft_gaseous.pdf?sfvrsn=4.

68. Columbia professor Andrea Baccarelli, for example, has shown that pollutants cause epigenetic damage through a process called DNA methylation, in which a molecule known as a *methyl group* attaches itself to DNA, silencing some genes and activating others. See Nancy Averett, "Scientists Leverage Big Data for Precision Public Health," *Columbia Public Health*, 2017, 20. Since 2015, Chinese scientists have also demonstrated the epigenetic impact of ambient PM2.5. See Renjie Chen, Liping Qiao, Huichu Li, et al., "Fine Particulate Matter Constituents, Nitric Oxide Synthase DNA Methylation and Exhaled Nitric Oxide," *Environmental Science & Technology* 49, no. 19 (September): 11859–11865. A more recent publication by Chinese scientists found that exposure to PM2.5 could significantly increase stress hormones in humans, as well as cause metabolic changes in blood glucose, amino acids and lipids. See Huichu Li, Jing Cai, Renjie Chen, et al., "Particulate Matter Exposure and Stress Hormone Levels: A Randomized, Double-Blind, Crossover Trial of Air Purification," *Circulation* 136, no. 7, 15 (August, 2017): 618–662.
69. Andrea A. Baccarelli, Nick Hales, Richard T. Burnett, et al., "Particulate Air Pollution, Exceptional Aging, and Rates of Centenarians: A Nationwide Analysis of the United States, 1980–2010," *Environmental Health Perspectives* 124, no. 11 (November, 2016): 1744.
70. Philip J. Landrigan and Mary M. Landrigan, *Children and Environmental Toxins: What Everyone Needs to Know* (New York, NY: Oxford University Press, 2018).
71. Philip J. Landrigran, et al., "The Lancet Commission on Pollution and Health," *The Lancet Commission*, October 19, 2017.
72. Frederica Perera, "Pollution from Fossil-Fuel Combustion Is the Leading Environmental Threat to Global Pediatric Health and Equity: Solutions Exist," *International Journal of Environmental Research and Public Health* 15,no. 1 (December, 2017): E16.
73. Jordan L. Schnell and Michael J. Prather, "Co-occurring Ozone, PM, and Temperature Extremes," *Proceedings of the National Academy of Sciences* 114, no. 11 (March, 2017): 2854–2859; Kai Chen, et al., "Does Temperature-Confounding Control Influence the Modifying Effect of Air Temperature in Ozone–Mortality Associations?" *Environmental Epidemiology* 2, no. 1 (March, 2018): e008.
74. World Health Organization, "7 Million Premature Deaths Annually Linked to Air Pollution," Geneva: WHO, March 25, 2014. www.who.int/mediacentre/news/releases/2014/air-pollution/en/; Landrigran, et al., "The Lancet Commission on Pollution and Health."
75. Damian Carrington and Matthew Taylor, "Air Pollution is the 'New Tobacco,' Warns WHO Head," *The Guardian*, October 27, 2018. www.theguardian.com/environment/2018/oct/27/air-pollution-is-the-new-tobacco-warns-who-head.
76. Michael Brauer, "Poor Air Quality Kills 5.5 Million Worldwide Annually," Institute for Health Metrics and Evaluation, February 12, 2016. www.healthdata.org/news-release/poor-air-quality-kills-55-million-worldwide-annually.
77. Rada Akbar, "Ten Threats to Global Health in 2019." The World Health Organization (2019). www.who.int/emergencies/ten-threats-to-global-health-in-2019.

78. Benjamin Bowe, Yan Xie, Yan Yan, Ziyad Al-Aly, "Burden of Cause-Specific Mortality Associated With PM2.5 Air Pollution in the United States," *JAMA Network Open*, 2 no. 11 (2019): e1915834.
79. C. Arden Pope III, Richard T. Burnett and Michael J. Thun, "Lung Cancer, Cardiop-ulmonary Mortality, and Long-Term Exposure to Fine Particulate Air Pollution," *JAMA* 287, 9 (March, 2002): 1132–1141.
80. Aaron J. Cohen, et al., "Estimates and 25-Year Trends of the Global Burden of Disease Attributable to Ambient Air Pollution: An Analysis of Data from the Global Burden of Diseases Study 2015," *The Lancet* 389, no. 10082 (May, 2017): 1912. www.thelancet.com/journals/lancet/article/PIIS0140-6736(17)30505-6/fulltext.
81. Benjamin Bowe, Yan Xie, Tingting Li, et al., "The 2016 Global and National Burden of Diabetes Mellitus Attributable to PM2·5 Air Pollution: A Longitudinal Cohort Study Using the Global Burden of Disease 2016 Data and Methodologies," *Lancet Planetary Health* (July, 2018): e301–e312.
82. Cohen, et al., "Estimates and 25-Year Trends of the Global Burden of Disease Attributable to Ambient Air Pollution: An Analysis of Data from the Global Burden of Diseases Study 2015."
83. Barbara A. Maher, Imad A. M. Ahmed, Vassil Karloukovski, et al., "Magnetite Pollution Particles in the Human Brain," *Proceedings of the National Academy of Sciences* 113, no. 39 (September, 2016): 10797–10801; Barbara A. Maher, David Allsop, Vassil Karloukovski, et al., "Toxic Air Pollution Nanoparticles Discovered in the Human Brain," Lancaster University, September 5, 2016. www.lancaster.ac.uk/news/articles/2016/toxic-air-pollution-nanoparticles-discovered-in-the-human-brain/.
84. Zen Vuong, "Air Pollution May Lead to Dementia in Older Women," USC News, January 31, 2017. https://news.usc.edu/115654/air-pollution-may-lead-to-dementia-in-older-women/.
85. National Center for Environmental Health, "Waterborne Illnesses," Vessel Sanitation Program, Atlanta, GA. www.cdc.gov/nceh/vsp/training/videos/transcripts/water.pdf.
86. Sana Ullah, Muhammad Javed, Muhammad Shafique, et al., "An Integrated Approach for Quality Assessment of Drinking Water Using GIS: A Case Study of Lower Dir," *Journal of Himalayan Earth Sciences* 47, no 2 (July, 2014): 163–174.
87. Robert D. Morris, "Drinking Water and Cancer," *Environmental health perspectives* 103, Suppl 8 (1995): 225–231; Sunderrajan Krishnan and Rajnarayan Indu, "Groundwater Contamination in India: Discussing Physical Processes, Health and Sociobehavioral Dimensions," IWMI-Tata, Water Policy Research Programmes, Anand, India (2006).
88. Sadia Jabeen, Quaisar Mehmood, Sumbal Tariq, et al., "Health Impact Caused by Poor Water and Sanitation in District Abbottabad," *Journal of Ayub Medical College* 23, no. 1 (January, 2011): 47–50.
89. Janet Currie, Joshua Graff Zivin, Katherine Meckel, et al., "Something in the Water: Contaminated Drinking Water and Infant Health," *Canadian Journal of Economics* 46, 3 (August): 791–810. A more recent study isolates the effect of lead from other chemicals

and finds the negative health effects are larger for black or less educated mothers. See Rui Wang, Xi Chen, Xun Li, "Something in the Pipe: Flint Water Crisis and Health at Birth," *IZA DP* No. 12115 (2019). www.iza.org/publications/dp/12115/something-in-the-pipe-flint-water-crisis-and-health-at-birth.

90. Science Communication Unit, University of the West of England, "Science for Environment Policy In-Depth Report: Soil Contamination: Impacts on Human Health," Report produced for the European Commission DG Environment, Issue 5, September 2013. http://ec.europa.eu/environment/integration/research/newsalert/pdf/IR5_en.pdf.
91. Science Communication Unit, University of the West of England, "Science for Environment Policy In-Depth Report: Soil Contamination."
92. "Why Young Children Are So Vulnerable to Lead," *New York Times*, November 1, 1990. www.nytimes.com/1990/11/01/us/why-young-children-are-so-vulnerable-to-lead.html; "Learn about Lead," Environmental Protection Agency, Washington, D.C. www.epa.gov/lead/learn-about-lead.
93. Landrigran, et al., "The Lancet Commission on Pollution and Health."
94. WHO, "Country Profile of Environmental Burden of Disease. China," Public Health and the Environment, Geneva (2009). www.who.int/quantifying_ehimpacts/national/countryprofile/china.pdf.
95. With 2.5 million deaths, India had the greatest number of pollution-related deaths. The Lancet, "The Lancet: Pollution Linked to Nine Million Deaths Worldwide in 2015, Equivalent to One in Six Deaths," *EutrkAlert!*, October 19, 2017. www.eurekalert.org/pub_releases/2017-10/tl-tlp101817.php.
96. Rohde and Muller, "Air Pollution in China: Mapping of Concentrations and Sources."
97. During 2005–2014, levels of NO_2 rose sharply over some cities, especially those in north central plain and dropped over others, including Shanghai, Guangzhou, and even Beijing. See Bryan N. Duncan, et al., "A Space-Based, High-Resolution View of Notable Changes in Urban NOx Pollution around the World (2005–2014)," *Journal of Geophysical Research Atmospheres*, 121 (January, 2016): 976–996. In January 2017, Linfen city of Shanxi Province, where PM2.5 and PM10 readings were often off the charts, saw the concentration of sulfur dioxide soar to 1,303 *ug*/m^3, nearly 18 times the national standard. Sun Ruisheng and Zheng Jinran, "Environment Experts Look into High Levels of Sulfur in Linfen," *China Daily*, January 13, 2017.
98. Hepeng Jia and Ling Wang, "Peering into China's Thick Haze of Air Pollution," *Chemical and Engineering News* 95, 4 (January, 2017): 19–22. https://cen.acs.org/articles/95/i4/Peering-Chinas-thick-haze-air.html.
99. Junfeng (Jim) Zhang and Jonathan M. Samet, "Chinese Haze versus Western Smog: Lessons Learned," *Journal of Thoracic Disease* 7, no. 1 (January, 2018): 3–13.
100. Guojia aizheng zhongxin (China National Cancer Center), "2018 nian quanguo zuixin aizheng baogao."

101. Qing Wang and Zhiming Yang, "Industrial Water Pollution, Water Environment Treatment, and Health Risks in China," *Environmental Pollution* 218 (November, 2016): 359.
102. Vivian C. Pun, Justin Manjourides and Helen Suh, "Association of Ambient Air Pollution with Depressive and Anxiety Symptoms in Older Adults: Results from the NSHAP Study," *Environmental Health Perspectives* 125 (March, 2017): 342–348.
103. Aaron J. Cohen, et al., "Estimates and 25-year Trends of the Global Burden of Disease Attributable to Ambient Air Pollution: An Analysis of Data from the Global Burden of Diseases Study 2015," *The Lancet* 389, no. 10082 (May, 2017): 1907–1918.
104. Die Fang, Qin'geng Wang, Huiming Li, et al., "Mortality Effects Assessment of Ambient PM2. 5 Pollution in the 74 Leading Cities of China," *Science of The Total Environment* 569 (July, 2016): 1545–1552.
105. Wanqing Chen, et al., "Disparities by Province, Age, and Sex in Site-Specific Cancer Burden Attributable to 23 Potentially Modifiable Risk Factors in China: a Comparative Risk Assessment," *Lancet Global Health* 7, no. 2 (February, 2019): e257–e269.
106. Renjie Chen, et al., "Fine Particulate Air Pollution and Daily Mortality. A Nationwide Analysis in 272 Chinese Cities," *American Journal of Respiratory and Critical Care Medicine* 196, no. 1 (July, 2017): 73–81.
107. "Shinian ertong xiaochuanbing zengjia 1.3 bei" ("Cases of Pediatric Asthma Increased 1.3 Fold in 10 Years,") *Souhu*, May 13, 2017. www.sohu.com/a/140381064_683361.
108. *Xiaochuan hangye yanjiu baogao* (Asthma industry research report), Sina, July 31, 2018. http://k.sina.com.cn/article_5334569296_13df7115002000acuh.html.
109. According to MEP, vehicle emission has become the no. 1 source of PM in many large and medium-sized cities by the end of 2016. It accounts for 31 percent of PM in Beijing, 29 percent in Shanghai, 28 percent in Hangzhou, and 22 percent in Guangzhou. See *Huanbaobu daqi wuran fangzhi meiti jianmianhui shilu* (Transcript of the MEP press conference on air pollution control), January 12, 2017. www.gov.cn/xinwen/2017-01/12/content_5159192.htm.
110. "Guowei fu xunshiyuan jiedu beifang diqu dongji qingjie qunuan guihua (2017–2021)" ("Deputy Inspector Guo Wei interprets Clean Energy Heating Plan for Winter in Northern Region 2017–2021"), *International Energy Information Network*, January 24, 2018. www.in-en.com/article/html/energy-2265139.shtml.
111. Yuyu Chen, et al., "Evidence on the Impact of Sustained Exposure to Air Pollution on Life Expectancy from China's Huai River Policy," *PNAS* 110, no. 32 (August, 2013): 12936–12941. www.pnas.org/content/110/32/12936.abstract.
112. John Kemp, "For Polluted China, Gas Is a Matter of Life or Death," *Reuters*, May 21, 2014. www.reuters.com/article/china-russia-gas-pollution/column-for-polluted-china-gas-is-a-matter-of-life-or-death-kemp-idUSL6N0O73HS20140521.
113. Avraham Ebenstein, "New Evidence on the Impact of Sustained Exposure to Air Pollution on Life Expectancy from China's Huai River Policy," *PNAS* 114, no. 39 (September, 2017): 10384–10389. www.pnas.org/content/early/2017/09/05/1616784114.full.

114. Li Jing, "Smog Could Shrink Chinese Economy by up to 2.6 Per cent by 2060: Study," *South China Morning Post*, June 10, 2016. www.scmp.com/news/china/policies-politics/article/1971971/smog-could-shrink-chinese-economy-26-cent-2060-study.
115. Fang, Wang, Li, et al., "Mortality Effects Assessment of Ambient PM2. 5 Pollution in the 74 Leading Cities of China."
116. "Cong Chai Jing jilupian li dudao de zhenxiang" ("Truth Found in Chai Jing's Documentary"), *Souhu*, February 28, 2015. http://business.sohu.com/20150228/n409230135.shtml.
117. "Woguo huocheng shijie diyi feiai daguo" ("Our Country Might Become the World's Lung Cancer Capital"), *Renmin ribao*, December 12, 2014. The stabilization of the growth of smoking-caused lung cancer incidence rate is likely due to the lack of significant change in China's overall smoking rate. See Gillian Wong, "China Tobacco Profits Undermine Anti-smoking Push," *MedicalXpress.com*, January 6, 2011. https://medicalxpress.com/news/2011-01-china-tobacco-profits-undermine-anti-smoking.html.
118. Xuexi Tie, Dui Wu, and Guy Brasseur, "Lung Cancer Mortality and Exposure to Atmospheric Aerosol Particles in Guangzhou, China," *Atmospheric Environment* 43 (May 2009): 2375–2377. www.researchgate.net/publication/222539592_Lung_cancer_mortality_and_exposure_to_atmospheric_aerosol_particles_in_Guangzhou_China.
119. "Diaocha xianshi 71.8% shoufangzhe ganjue zheng shou shuiwuran weixie" ("Survey Suggests that 71.8% of Respondents Feel Threatened by Water Pollution"), *Zhongguo qingnian bao* (*China Youth Daily*), February 7, 2013, 7. http://finance.sina.com.cn/china/20130207/071414527255.shtml.
120. Jane Qiu, "China to Spend Billions Cleaning Up Groundwater," *American Association for the Advancement of Science* 334, no. 6057 (November, 2011): 745. http://science.sciencemag.org/content/334/6057/745.
121. Huang, *Governing Health in Contemporary China*.
122. Nian-Feng Lin, Jie Tang, Hoteyi S. M. Ismael, "Study on Environmental Etiology of High Incidence Areas of Liver Cancer in China," *World Journal of Gastroenterology* 6 (September, 2000): 572–576; Avraham Ebenstein, "The Consequences of Industrialization: Evidence from Water Pollution and Digestive Cancers in China," *Review of Economic Statistics* 94, MIT Press (February, 2012): 186–201; Yang Gao, et al., "Groundwater Nitrogen Pollution and Assessment of Its Health Risks: a Case Study of a Typical Village in Rural–Urban Continuum, China," *PLOS One* 7, no. 4 (April, 2012): e33982.; Hongwei Sun, et al., "Distribution, Possible Sources, and Health Risk Assessment of SVOC Pollution in Small Streams in Pearl River Delta, China," *Environ. Sci. Pollut. Res.* 21, no. 17 (September, 2014): 10083–10095.
123. Gonghuan Yang and Dafang Zhuang (eds.), *Atlas of the Huai River Basin Water Environment: Digestive Cancer Mortality* (Netherlands: Springer, 2014)

124. Xueyan Zhang, et al., "Esophageal Cancer Spatial and Correlation Analyses: Water Pollution, Mortality Rates, and Safe Buffer Distances in China," *Journal of Geographical Sciences* 24, no. 1 (December, 2013): 46–58.
125. Qing Wang and Zhiming Yang, "Industrial Water Pollution, Water Environment Treatment, and Health Risks in China," *Environmental Pollution* 218 (November, 2016): 258–365.
126. Jack E. Fergusson, *The Heavy Elements: Chemistry, Environmental Impact, and Health Effects,* First Edition (Oxford, England: Pergamon Press, 1990).
127. In 1974, Chinese scientists found that rice in Shenyang was contaminated by cadmium due to irrigation using polluted water. See "Buan de liangshi zhongjinshu anquan" ("Worrisome Food Heavy Metal Safety"), *Dongfang zaobao* (*Oriental Morning Post*), August 20, 2013. http://fashion.ifeng.com/health/pingmei/detail_2013_08/20/28810906_0.shtml.
128. Yonglong Lu, et al., "Impacts of Soil and Water Pollution on Food Safety and Health Risks in China," *Environment International* 77 (April, 2015): 11.
129. Chen Nengchang, "Ying keguan kandai nongtian ge wuran" ("Farmland Cadmium Contamination Should Be Viewed By Objectively"), *Zhongguo kexue bao* (*China Science Daily*), August 9, 2017. http://news.sciencenet.cn/htmlnews/2017/8/384634.shtm.
130. Ministry of Environmental Protection and Ministry of Land and Resources, *Quanguo turang wuran zhuangkuang diaocha gongbao.*
131. Lv Yuhua, *Woguo ertong xueqian shuiping xianzhuang ji duice yanjiu* (*A Study of BLL Situation and Solutions in Our Country*), Master's Thesis, Nanhua University, China (2014).
132. Kangmin He, Shunqin Wang, and Jinliang Zhang, "Blood Lead Levels of Children and Its Trend in China," *Science of the Total Environment* 407, no. 13 (June 2009): 3986–3993.
133. Tanmin Gao Junquan, "Xueqian: shenxiang ertong jiankang de mozhao" ("Blood Lead: Devil's Talons Extended to Child Health"), *Lvse jiayuan* (*Green Homeland*), no. 12 (December, 2003). www.people.com.cn/GB/paper2742/11239/1015892.html.
134. Andri Bryner, "New Risk Model Sheds Light on Arsenic Risk in China's Groundwater," *Eawag,* August 22, 2013. www.eawag.ch/en/news-agenda/news-portal/news-archive/archive-detail/news/neues-modell-zeigt-arsenrisiko-aus-chinas-grundwasser/.
135. Peter Muennig, Yue Wang, and Aleksandra Jakubowski, "The Health of Immigrants to New York City from Mainland China: Evidence From the New York Health Examination and Nutrition Survey," *Journal of Immigrant & Refugee Studies* 10 (January, 2012): 131–137.
136. Yonglong Lu, et al., "Impacts of Soil and Water Pollution on Food Safety and Health Risks in China."
137. Yanzhong Huang, "Are China's Chickens Contaminating America's Plates?" *Foreign Policy,* November 16, 2017. https://foreignpolicy.com/2017/11/16/are-chinas-chickens-contaminating-americas-plates/.
138. David Stanway, "China Uncovers 500,000 Food Safety Violations in Nine Months," *Reuters,* December 24, 2016. www.reuters.com/article/us-china-food-safety/china-uncovers-500000-food-safety-violations-in-nine-months-idUSKBN14D046.

139. M. Anju and D.K. Banerjee, "Multivariate Statistical Analysis of Heavy Metals in Soils of a PbeZn Mining Area, India" *Environmental Monitoring and Assessment* 184, no. 7 (2012): 4191–4206; Yang-Guang Gu, Qu Sheng Li, Jian Hong Fang, et al., "Identification of Heavy Metal Sources in the Reclaimed Farmland Soils of the Pearl River Estuary in China Using a Multivariate Geostatistical Approach," *Ecotoxicology and Environmental Safety* 105 (April, 2014): 7–12; Wen-Yan Han, Fang-Jie Zhao, Yuan-Zhi Shi, et al., "Scale and Causes of Lead Contamination in Chinese Tea," *Environmental Pollution* 139 (January, 2006): 125–132.
140. "Meinian zhongjinshu wuran 1200 wan dun liangshi" ("Heavy Metal Contaminates 12 Million Tons of Grain Each Year"), *Diyi caijing (yicai)*, February 7, 2013. www.yicai.com/news/2481043.html.
141. Qihong Zhao, et al., "Potential Health Risks of Heavy Metals in Cultivated Topsoil and Grain, Including Correlations with Human Primary Liver, Lung and Gastric Cancer, in Anhui Province, Eastern China," *Science of the Total Environment* 470–471 (Febuary, 2014): 340–347.
142. Ying Huang, Qianqian Chen, Meihua Deng, et al., "Heavy Metal Pollution and Health Risk Assessment of Agricultural Soils in a Typical Peri-urban Area in Southeast China," *Journal of Environmental Management* 207 (February, 2018): 159–168.
143. Liang Wei, "Meinian zhongjinshu wuran 1200 wandun liangshi" ("12 Million Tons of Grain are Contaminated Annually by Heavy Metals) *Shidai zhoubao* (*Time Weekly*), February 7, 2013. http://money.163.com/13/0207/06/8N3D2KQN00253B0H.html.
144. Yan Song, Yibana Wang, Weifeng Mao, et al., "Dietary cadmium exposure assessment among the Chinese population," *PLOS ONE* 12, no. 5 (May, 2017): e0177978. https://doi.org/10.1371/journal.pone.0177978.
145. Shang Qi, Zhai Miaomiao, Yao Liangsan, et al., "Jiangxi sheng mouxian dami ge wuran qingkuang zhuizong diaocha" ("A Tracking Study on Cadmium Contamination of Rice in One County of Jiangxi Province"), *Weisheng yanjiu* (*Journal of Hygiene Research*) 38, no. 3 (May, 2009): 296–298; Wen-Li Zhang, Yu Du, Miao-Miao Zhai, et al., "Cadmium Exposure and Its Health Effects: A 19-year Follow-Up Study of a Polluted Area in China," *Science of The Total Environment*, 470–471 (February, 2014): 224–228.
146. Song, Wang, Mao, et al., "Dietary Cadmium Exposure Assessment among the Chinese Population."
147. Yanzhong Huang, "China's Worsening Food Safety Crisis," *The Atlantic*, August 28, 2012. www.theatlantic.com/international/archive/2012/08/chinas-worsening-food-safety-crisis/261656/.
148. Pan Jie, "Du dami bu du naifen houchen" ("Toxic Rice Follows in Footsteps of Toxic Milk Powder"), *Renmin wang*, May 28, 2013. http://paper.people.com.cn/gjjrb/html/2013-05/28/content_1245937.htm.
149. "Hunan Zhuzhou guafucun wuran sinue" ("Pollution is a Big Problem in the 'Widow Village' in Zhuzhou, Hunan"), *Yangguang wang*, January 1, 2015. http://news.cnr.cn/native/pic/20150122/t20150122_517493259.shtml.
150. Sun Xuemei, Wang Shuo, Niu Yinghui, "Zhongguo yin wuran suo chansheng aizheng-cun bei zhi xiang zhongxibu zhuanyi" ("Cancer Villages Caused by Pollution are Noted

to be Moving toward Central West China"), *Jinghua shibao* (*Beijing Times*), February 22, 2013. http://green.sina.com.cn/news/roll/2013-02-22/104926329327.shtml; Gao Changan, "Wu Yiling: Chongfen zhongshi aizhengcun deng huanjing wuran xianxiang pinfa wenti" ("Pay Adequate Attention to Frequent Occurrence of Cancer Villages and Other Environmental Pollution Problems"), *ScienceNet.com*, March 3, 2017. http://news.sciencenet.cn/htmlnews/2017/3/369429.shtm.

151. Deng Fei, "Zhongguo baichu zhiai weidi" ("A Hundred Cancer-causing Dangerous Places in China"), *Fenghuang zhoukan* (*Phoenix Weekly*), no. 11 (April, 2009).
152. "Zhongguo aizhengcun ditu jiemi" ("Deciphering China's Cancer Village Map"), *Renmin Wang*, February 25, 2013. http://jiankang.cntv.cn/2013/02/25/ARTI1361754058324182.shtml.
153. Lee Liu, "Made in China: Cancer Villages," *Environment*, March/April 2010. www.environmentmagazine.org/Archives/Back%20Issues/March-April%202010/made-in-china-full.html.
154. CDC, "Cancer Clusters," *National Center for Environmental Health*, December 18, 2013. www.cdc.gov/nceh/clusters/; Sheng-Sheng Gong and Tao Zhang, "Temporal-spatial Distribution Changes of Cancer Villages in China," *China Population, Resources, Environment* 23, no. 9 (2013), 156–164.
155. Yao Shuxiang et al., "Yunnan xikuang faiai gaofa de liuxingbingxue diaocha, 1954–2002" ("Epidemiologic Investigation of Occupational Lung Cancer in Yunnan Tin Miners During 1954–2002"), *Huanjing yu Zhiye Yixue* (*Journal of Environmental & Occupational Medicine*) 24, no. 5 (October 2007): 465–468.
156. Yonglong Lu, et al., "Impacts of Soil and Water Pollution on Food Safety and Health Risks in China."
157. Gong and Zhang, "Temporal-Spatial Distribution Changes of Cancer Villages in China."
158. Gong and Zhang, "Temporal-Spatial Distribution Changes of Cancer Villages in China."
159. Gong and Zhang, "Temporal-Spatial Distribution Changes of Cancer Villages in China."
160. Environmental scholars at Nanjing University, interview with author, Nanjing, June 29, 2015.
161. Ebenstein, "The Consequences of Industrialization: Evidence from Water Pollution and Digestive Cancers in China."
162. Professor Ran Tao, Renmin University, China, Communication with author, June 23, 2017.
163. Dan Zheng and Minjun Shi, "Multiple Environmental Policies and Pollution Haven Hypothesis: Evidence from China's Polluting Industries," *Journal of Cleaner Production* 141 (January, 2017): 299.
164. Anna Lora-Wainwright, *Fighting for Breath: Living Morally and Dying of Cancer in a Chinese Village* (University of Hawai'i Press, 2013).
165. Li Caihong and Cheng Pengli, "Wenti hua yu qu wuming hua" (Problematization and de-stigmatization), in Chen Ajiang, Cheng Pengli and Luo Yajuan (eds.), *Aizheng cun*

diaocha (Investigation of Cancer Villages) (Beijing: Shehui kexue chubanshe, 2013), 121–141.

166. See, for example, "Nuer yin shuiwuran huan baixuebing qushi nongmin shinian xun zhenxiang gao huanbaobu" ("After His Daughter Died of Water-pollution Caused Leukemia, a Farmer Spent Ten Years Seeking Truth and Suing Ministry of Environmental Protection"), *Beijing qingnian bao* (*Beijing Youth Daily*), November 17, 2016.
167. Benjamin Van Rooij, "The People vs. Pollution: Understanding Citizen Action against Pollution in China," *The Journal of Contemporary China* 19, no. 63 (January, 2010): 55–79.
168. Professor Yang Gonghuan, Beijing, interview with author, July 2, 2015; Li Hujun, "Jujiao huaihe xiayou aizhengcun" ("Focusing on Cancer Villages in the Lower Reaches of Huai River"), *Caijing,* no. 14 (July, 2008). http://magazine.caijing.com.cn/2008-07-05/110066241.html.
169. Zhu Wenhe, the Center for Legal Assistance to Pollution Victims, Beijing, interview with author, July 3, 2015.
170. Jennifer Holdaway and Wang Wuyi, "Preface" in *Aizheng cun diaocha,* 14.
171. Professor Yang Gonghuan, Beijing, interview with author, July 2, 2015.
172. Zhang Zhilan, et al., "Yangzhongxian exing zhongliu siwang qingkuang diaocha yanjiu (1973–1982)" ("Investigation and Research of Malignant Cancer Mortality in Yangzhong County, 1973–1982"), *Zhonghua zhongliu zazhi* (*Chinese Journal of Oncology*) 10, no. 2 (1988): 102–104; Cai Dehua, "Yangzhongxian turang beijingzhi jiqi yu exing zhongliu siwanglu guanxi" ("The Correlation between Soil Background Value and Malignant Cancer Mortality Rate in Yangzhong County"), *Shengtai yu nongcun huanjing xuebao* (*Ecology and Rural Environment Journal*), no. 2 (1993); Guo Guoping and Hua Zhaolai, "Yangzhong shi butong shiqi exing zhongliu fabing tongji fenxi" ("A Statistical Analysis of Malignant Cancer Incidences in Different Periods of Time in Yangzhong City"), *Zhongguo zhongliu* (*China Oncology*), 13, no. 3 (2004).
173. Phil Brown, *Toxic Exposures: Contested Illness and the Environmental Health Movement* (New York, NY: Columbia University Press, 2007); Sylvia Noble Tesh, *Uncertain Hazards: Environmental Activists and Scientific Proof* (Ithaca, NY: Cornell University Press, 2000).
174. Holdaway and Wang, "Preface."
175. Javier C. Hernandez, "'No Such Thing as Justice' in Fight Over Chemical Pollution in China," *New York Times,* June 12, 2007. www.nytimes.com/2017/06/12/world/asia/china-environmental-pollution-chemicals-lead-poisoning.html.
176. Yuan Suwen and Du Caicai, "Xiaoshi de fengjun he wuran shouhaizhe jiuzhu kunjing" ("The 'Missing' Feng Jun and the Pollution Victims' Assistance Dilemma"), *Caixin zhoukan* (*Caixin Weekly*), October 2, 2017. http://china.caixin.com/2017-10-02/101153054.html.
177. Dan Hoornweg, Philip Lam, and Manisha Chaudhry, "Waste Management in China: Issues and Recommendations," Urban Development Working Papers; no. 9. Washington, DC: World Bank, 2005. http://documents.worldbank.org/curated/en/237151468025135801/Waste-management-in-China-issues-and-recommendations

178. "Laji weicheng: bi wumai geng chumujingxin" ("Being Besieged by Garbage is Even More Shocking by the Sight than Smog"), *Shijie huaren zhoukan (World Chinese Weekly)*, January 5, 2017. http://wemedia.ifeng.com/6772733/wemedia.shtml.
179. "Wangbao Beijing yibei laji qihuan weicheng" ("It Was Reported Online that Beijing Has Been Encircled by 'The Seventh Ring Garbage'"), *Guangzhou ribao* (*Guangzhou Daily*), June 27, 2011. http://media.people.com.cn/GB/40606/15000721.html.
180. "Laji weicheng: bi wumai geng chumujingxin" ("Being Besieged by Garbage is Even more Shocking by the Sight than Smog).
181. Lin Jie, "Woguo 1/3 chengshi bei laji baowei" ("1/3 of the Cities in Our Country are Besieged by Garbage"), *Yangcheng wanbao* (*Yangcheng Evening News*), October 14, 2013. http://news.sina.com.cn/c/2013-10-14/064928425686.shtml.
182. Li He, "Zhongguo sanfen zhier dazhong chenshi xian laji baowei" ("Two Thirds of China's Large and Medium-sized Citeis are Besieged by Waste"), *Keji ribao* (*Science and Technology Daily*), February 13, 2018. http://finance.sina.com.cn/roll/2018-02-13/doc-ifyrkrva8208131.shtml.
183. Chen Boyuan, "Garbage Besieges One Third of Chinese Cities," *China.org.cn*, July 19, 2013. www.china.org.cn/environment/2013-07/19/content_29473726.htm.
184. Ye Zhou, "Woguo meinian chansheng jin shiyi dun laji" ("Our Country Produces One Billion Tons of Garbage Annually"), *Jinghua shibao* (*Beijing Times*), July 8, 2011. http://news.sina.com.cn/green/2011-07-08/131422780229.shtml.
185. The National Bureau of Statistics of China, *Zhongguo tongji nianjian 2016* (*China Statistical Yearbook 2016*) (Beijing: Zhongguo tongji chubanshe, 2016).
186. "70 Percent of Annual Global E-waste Dumped in China," *Beijing Review*, May 24, 2012. www.bjreview.com.cn/Energy/txt/2012-05/24/content_454772.htm.
187. He Tao, Xi Nan and He Guangwei, "Laji zhen bianshen" ("The transformation of 'Garbage Town'"), *Caijing* (*Finance and Economics*), August 3, 2017. http://m.caijing.com.cn/article/116857.
188. He, Xi, and He, "Laji zhen bianshen" ("The Transformation of 'Garbage Town'").
189. "Guiyu: An E-Waste Nightmare," *Greenpeace*. www.greenpeace.org/eastasia/campaigns/toxics/problems/e-waste/guiyu/.
190. Jiasaki Guogang, "Zhiji Beijing zuichou laji tianmaichang" ("First-hand Witness of the Most Stinky Refuse Landfill"), *Duowei News*, August 12, 2017. http://news.dwnews.com/china/photo/2017-08-12/60006518.html?p=14.
191. "Guiyu: An E-Waste Nightmare."
192. Dongliang Zhang, Guangqing Huang, Yimin Xu, et al., "Waste-to-Energy in China: Key Challenges and Opportunities," *Energies* 8 no. 12 (December, 2015): 14182–14196. https://pdfs.semanticscholar.org/9224/0740d8a71c601d7956a10ebd1ecc54564bbe.pdf.
193. "Dioxins and Their Effects on Human Health," World Health Organization, October 4, 2016. www.who.int/en/news-room/fact-sheets/detail/dioxins-and-their-effects-on-human-health.
194. Ni Yuwen, Haijun Zhang, Su Fan, et al., "Emissions of PCDD/Fs from Municipal Solid Waste Incinerators in China," *Chemosphere* 75 (April, 2009): 1153–1158.

195. Guogang Jiaqi, "Zhongguo laji weicheng zhi shang" ("The Sad Story of Chinese Cities Besieged by Garbage"), *Duowei News*, August 22, 2017. http://news.dwnews.com/china/news/2017-08-22/60008269_all.html.
196. Urszula Łopuszańska and Marta Makara-Studzińska, "The Correlations between Air Pollution and Depression," *Current Problems of Psychiatry* 18, 2 (September, 2017): 100–109; Mieczysław Szyszkowicz, Jeff B. Willey, Eric Grafstein, et al., "Air Pollution and Emergency Department Visits for Suicide Attempts in Vancouver, Canada," *Environmental Health Insights* 4 (October, 2010): 79–86; Vivian C. Pun, Justin Manjourides, Helen Suh, "Association of Ambient Air Pollution with Depressive and Anxiety Symptoms in Older Adults: Results from the NSHAP Study." *Environmental Health Perspect* 125,no. 3 (March, 2017): 342–348; Youngdon Kim, Woojae Myung, Hong-Hee Won, et al., "Association between Air Pollution and Suicide in South Korea: A Nationwide Study," *PLOS ONE* (February, 2015). https://doi.org/10.1371/journal.pone.0117929; Amanda V. Bakian, Rebekah S. Huber, Hilary Coon, et al., "Acute Air Pollution Exposure and Risk of Suicide Completion," *American Journal of Epidemiology* 181, no. 5 (March, 2015): 295–303. https://doi.org/10.1093/aje/kwu341; Atif Khan, Oleguer Plana-Ripoll, Sussie Antonsen, et al., "Environmental Pollution is Associated with Increased Risk of Psychiatric Disorders in the US and Denmark," *PLOS Biology* 17, no. 10 (August, 2019): e3000513.
197. "Ruhe shuaidiao wumai daigei nide xinli jiaolv" ("How to Get Rid of the Anxiety Brought by Smog"), *Fenghuang jiankang* (Phoenix Health), December 13, 2013. http://fashion.ifeng.com/health/news/dujiajianwen/wu/detail_2013_12/13/32114481_0.shtml.
198. Xin Zhang, Xiaobo Zhang and Xi Chen, "Happiness in the Air: How Does a Dirty Sky Affect Mental Health and Subjective Well-being?" *Journal of Environmental Economics and Management* 85 (September, 2017): 81–94. See also Siqi Zheng, et al., "Air Pollution Lowers Chinese Urbanites' Expressed Happiness on Social Media," *Nature Human Behavior* 3, no. 3 (January, 2019): 237–243.
199. Xin Zhang, Xi Chen, Xiaobo Zhang, "The Effects of Exposure to Air Pollution on Subjective Well-being in China," IZA DP No. 12313. In *Handbook on Well-being, Happiness, and the Environment* edited by D. Maddison, K. Rehdanz, H. Welsch. Dutch Post Foundation, 2019. www.iza.org/publications/dp/12313/the-effects-of-exposure-to-air-pollution-on-subjective-well-being-in-china.
200. Guo-Zhen Lin, Li Li, Yun-Feng Song, et al., "The Impact of Ambient Air Pollution on Suicide Mortality: a Case-Crossover Study in Guangzhou, China," *Environmental Health* 15 (August, 2016): 90.
201. Weina Liu, Hong Sun, et al., "Air Pollution Associated with Non-suicidal Self-Injury in Chinese Adolescent Students: A Cross-Sectional Study," *Chemosphere* 209 (October, 2018): 944–949.
202. Laura He, "China's First Listed Psychiatric Hospital Set to Break Mental Health Taboo, and Raise HK$681 Million in the Process," *South China Morning Post*, November 20, 2015. www.scmp.com/business/china-business/article/1880563/chinas-first-listed-psychiatric-hospital-set-break-mental.

203. Tom Philips, "Pollution Pushes Shanghai towards Semen Crisis," *The Telegraph*, November 7, 2013. www.telegraph.co.uk/news/worldnews/asia/china/10432226/Pollution-pushes-Shanghai-towards-semen-crisis.html.
204. Sun Wenyu, "Environmental Pollution One Cause of Declining Fertility in China: Expert," *People's Daily Online*, December 7, 2016. http://en.people.cn/n3/2016/1207/c90000-9151912.html.
205. "Woguo buyun buyulv da 12.5–15%" ("The Infertility Rate in Our Country Has Reached 12.5–15 Percent"), *Xin jing bao* (*The Beijing News*), October 30, 2018. www.bjnews.com.cn/health/2018/10/30/515645.html.
206. Dandan Li and Yinan Zhao, "China's Debate over a Shrinking Birth Rate Highlights Growth Concerns," *Bloomberg News*, January 12, 2019. www.bloomberg.com/news/articles/2019-01-03/china-debate-over-shrinking-birth-rate-highlights-growth-concern.
207. Liu Sha, "High Levels of Antibiotics in China's Rivers," *Global Times*, December 26, 2014. www.globaltimes.cn/content/898705.shtml; Chengcheng Jiang, "When Penicillin Pays: Why China Loves Antibiotics a Little Too Much," *Time*, January 5, 2012. http://content.time.com/time/world/article/0,8599,2103733,00.html.
208. Wang Dan, et al., "Zhongguo dibiaoshui huanjing zhong yaowu he geren hulipin de yanjiu jinzhan" ("The Progress of Research on the Drugs and Personal Care Products in China's Surface Water"), *Kexue tongbao* (*Chinese Science Bulletin*) 59, no. 743 (January, 2014).
209. Ke-Qing Xiao, Bing Li, Liping Ma, et al., "Metagenomic Profiles of Antibiotic Resistance Genes in Paddy Soils from South China," *FEMS Microbiology Ecology* 92, no. 3 (March, 2016).
210. Zhuang Pinghui, "Scientists Find Bacteria in Beijing Smog that Lead to Antibiotic Resistance," *South China Morning Post*, November 26, 2016. www.scmp.com/news/china/policies-politics/article/2049206/scientists-identify-bacteria-beijing-smog-lead.
211. Zhang Hongguang, Zheng Lei, "Wumai yanzhong, ta zhichu le wenti suozai, da'an rang women gandao jue wang" ("The Smog is Serious, and He Pinpoints the Problem, But the Answer Makes us Feel Desperate"), *Sina*, October 6, 2016. http://blog.sina.com.cn/s/blog_52f526870102x305.html.

ECONOMIC, SOCIOPOLITICAL, AND FOREIGN POLICY IMPACTS

1. David Roberts, "Underground Environmentalism in Communist East Germany," *Grist*, April 21, 2011. http://grist.org/article/2011–04-21-underground-environmentalism-in-communist-east-germany/.
2. Elizabeth Economy, *The River Runs Black: The Environmental Challenge to China's Future* (Ithaca, NY: Cornell University Press, 2004), 223.
3. Revolution Mauer and the Fall of the Wall, "The Opposition Goes Public." http://revolution89.de/en/awakening/the-opposition-goes-public/.
4. The Collective, "3 Top Concerns for Chinese Citizens in 2017," Collective Responsibility, January 24, 2017. www.coresponsibility.com/top-concerns-chinese-davos/.

5. Wang Tao and Zhu Hongjun "Sishi nian huanbao lu" ("Forty Years of Environmental Protection Path"), *Nanfang zhoumo* (*Southern Weekend*), June 9, 2013. www.infzm.com/content/91241.
6. Robert W. Fogel, "Health, Nutrition, and Economic Growth," *Economic Development and Cultural Change* 52, no. 3 (April, 2004): 643–658; Jeffrey Sachs, *Macroeconomics and Health: Investing in health for economic development,* Report of the Commission on Macroeconomics and Health Geneva: World Health Organization, 2001; T. Paul Schultz, "Health Human Capital and Economic Development," *Journal of African Economies* 19, no. 3 (November, 2010): iii12–iii80.
7. Sachs, *Macroeconomics and Health.*
8. Philip J. Landrigan, Richard Fuller, Nereus Acosta, et al., "Lancet Commission on Pollution and Health," *The Lancet Commissions* 391, no. 10119 (2018): 5.
9. Thomas D. Crocker and Robert L. Horst, Jr, "Hours of Work, Labor Productivity, and Environmental Conditions: A Case Study," *Review of Economics and Statistics,* MIT Press, 63, no. 3 (August, 1981): 361–368; Bart D. Ostro, "The Effects of Air Pollution on Work Loss and Morbidity," *Journal of Environmental Economics and Management* 10, no. 4 (December, 1983): 371–382; Jerry A. Hausman, Bart D. Ostro, and David A. Wise, "Air Pollution and Lost Work," National Bureau of Economic Research Working Paper No.1263 (1984); Richard T. Carson, Phoebe Koundouri, and Celine Nauges, "Arsenic Mitigation in Bangladesh: A Household Labor Market Approach," *American Journal of Agricultural Economics* 93, no. 2 (January, 2011): 407–414; Rema Hanna and Paulina Oliva, "The Effect of Pollution on Labor Supply: Evidence from a Natural Experiment in Mexico City," The National Bureau of Economic Research Working Paper No. 17302 (2011); Joshua Graff Zivin and Matthew Neidell, "The Impact of Pollution on Worker Productivity," *American Economic Review* 102, no. 7 (December, 2012): 3652–3673; Tom Chang, Joshua Graff Zivin, Tal Gross, et. al., "Particulate pollution and the productivity of pear packers," Working Paper No. 19944, National Bureau of Economic Research (2014); Tom Chang, Joshua Graff Zivin, Tal Gross, et al., "The Effect of Pollution on Worker Productivity: Evidence from Call-Center Workers in China," NBER Working Paper No. 22328 (June, 2016); Hong Soo Chew, Wei Huang and Xun Li, "Does Haze Cloud Decision Making? A Natural Laboratory Experiment," (January, 2018). https://ssrn.com/abstract=3102263.
10. John C. Ham, Jacqueline S. Zweig, and Edward Avol, "Pollution, Test Scores and the Distribution of Academic Achievement: Evidence from California Schools 2002–2008," IZA Conference Paper (October, 2011); Avraham Ebenstein, Victor Lavy, and Sefi Roth, "The Long-Run Economic Consequences of High-Stakes Examinations: Evidence from Transitory Variation in Pollution," *American Economic Journal: Applied Economics* 8 no. 4 (2016): 36–65.
11. Anna Aizer and Janet Currie, "Lead and Juvenile Delinquency: New Evidence from Liked Birth, School and Juvenile Detention Records," NBER Working Paper No. 23392 (May, 2017).

12. Xi Chen, "Smog, Cognition and Real-World Decision-Making," *International Journal of Health Policy and Management* 8, no. 2 (February, 2019): 76–80. www.ncbi.nlm.nih.gov/pmc/articles/PMC6462201/
13. Wu Xianhua, *Wumai wuran paifang de yingxiang jiqi guankong youhua* (*The Impact of Smog Pollution Emission and Its Optimal Control*) (Beijing: Kexue chubanshe, 2018).
14. Panle Jia Barwick, Shanjun Li and Deyu Rao, et al., "Air Pollution, Health Spending and Willingness to Pay for Clean Air in China," (July, 2017).
15. Xi Chen, Xiaobo Zhang and Xin Zhang, "The Impact of Exposure to Air Pollution on Cognitive Performance," *Proceedings of the National Academy of Sciences* 115, no. 37 (August, 2018): 9193–9197.
16. Yang Xia, et al., "Assessment of Socioeconomic Costs to China's Air Pollution," *Atmospheric Environment* 139 (August, 2016): 147–156.
17. Teng Li, Haoming Liu, and Alberto Salvo, "Severe Air Pollution and Labor Productivity," Department of Economics, National University of Singapore, April 10, 2015. http://conference.iza.org/conference_files/environ_2015/liu_h13306.pdf.
18. World Bank and Development Research Center of the State Council, *People's Republic of China, China 2030: Building a Modern, Harmonious, and Creative High-Income Society* (Washington, DC: World Bank, 2013), 39.
19. Keith Crane and Zhimin Mao, *Costs of Selected Policies to Address Air Pollution in China* (Santa Monica, CA: RAND Corporation, 2015), 3. www.rand.org/pubs/research_reports/RR861.html.
20. World Bank, *Cost of Pollution in China: Economic Estimates of Physical Damages* (Washington, DC: World Bank, 2007). http://documents.worldbank.org/curated/en/782171468027560055/Cost-of-pollution-in-China-economic-estimates-of-physical-damages.
21. Landrigan, Fuller, Acosta, et al., "The Lancet Commission on Pollution and Health."
22. "China Is Set to Lose 2% of GDP Cleaning Up Pollution," *Bloomberg Business*, September 17, 2010.
23. Xin Zhang, Xiaobo Chang, and Xi Chen, "Valuing Air Quality Using Happiness Data: The Case of China," *Ecological Economics* 137 (July, 2017): 29–36.
24. World Bank, *Cost of Pollution in China.*
25. World Bank, *Cost of Pollution in China.*
26. World Bank, *The Cost of Air Pollution: Strengthening the Economic Case for Action* (Washington, DC: World Bank Group, 2016). http://documents.worldbank.org/curated/en/781521473177013155/The-cost-of-air-pollution-strengthening-the-economic-case-for-action.
27. World Bank, *The Cost of Air Pollution,* 93.
28. Studies suggest that the morbidity costs resulting from pollution-related disease might conservatively increase mortality costs by 22–78 percent for China. See Landrigan, Fuller, Acosta, et al., "The Lancet Commission on Pollution and Health," p. 22.
29. Cao Caihong and Han Liyan, "Wumai dailai de shehui jiankang chengbeng gusuan" ("An Assessment of Smog-Incurred Social Health Cost"), *Tongji yanjiu* (*Statistical Research*), no. 7 (2015): 19–23.

30. Anjia Zheng, "Twice As Many Expatriates Leaving China Than Arriving, Moving Company Says," *Wall Street Journal*, February 9, 2015. https://blogs.wsj.com/chinarealtime/2015/02/09/twice-as-many-expatriates-leaving-china-than-arriving-moving-company-says/.
31. AmCham China, *China Business Climate Survey Report*, The American Chamber of Commerce in the People's Republic of China (2016), 26.
32. AmCham China, *China Business Climate Survey Report*, The American Chamber of Commerce in the People's Republic of China (2017, 2018, 2019).
33. AmCham China, *China Business Climate Survey Report*.
34. Li Dabai, "Ding Xueliang: muqian dalu meiyou chansheng guoji jinrong zhongxin de jiben tiaojian" ("Ding Xueliang: Mainland China Lacks Basic Requirements to Produce a Global Financial Center"), *ThinkerBig*, March 18, 2016. https://cul.qq.com/a/20160318/009960.htm.
35. UNECE, "Air Pollution and Food Production." www.unece.org/environmental-policy/conventions/envlrtapwelcome/cross-sectoral-linkages/air-pollution-and-food-production.html.
36. See World Bank, *Cost of Pollution in China*.
37. Xiaoping Wang and Denise Mauzerall, "Characterizing Distributions of Surface Ozone. and its Impact on Grain Production in China, Japan and South Korea: 1990 and 2020," *Atmospheric Environment* 38 (March, 2004): 4383–4402; Shiri Avnery, Denise L. Mauzerall, Junfeng Liu, et al., "Global Crop Yield Reductions Due to Surface Ozone Exposure: 2. Year 2030 Potential Crop Production Losses and Economic Damage under Two Scenarios of O3 Pollution," *Atmospheric Environment* 45, no. 13 (April 2011): 2297–2309.
38. Y. Gu, et al., "Impacts of Sectoral Emissions in China and the Implications: Air Quality, Public Health, Crop Production, and Economic Costs," *Environmental Research Letters* 13 (July, 2018): 1–13.
39. Jonathan Kaiman, "China's Toxic Air Pollution Resembles Nuclear Winter, Say Scientists," *The Guardian*, February 25, 2014. www.theguardian.com/world/2014/feb/25/china-toxic-air-pollution-nuclear-winter-scientists.
40. Li Jing, "Smog Could Shrink Chinese Economy by up to 2.6 Per cent by 2060: Study," *South China Morning Post*, June 10, 2016. www.scmp.com/news/china/policies-politics/article/1971971/smog-could-shrink-chinese-economy-26-cent-2060-study.
41. "China Is Set to Lose 2% of GDP Cleaning Up Pollution," *Bloomberg Business*, September 17, 2010.
42. "China Is Set to Lose 2% of GDP Cleaning Up Pollution."
43. Qing Wang and Zhiming Yang, "Industrial Water Pollution, Water Environment Treatment, and Health Risks in China," *Environmental Pollution* (November, 2016): 358–365.
44. Matthew E. Kahn, "The Link between Income Inequality and Exposure to Air Pollution in China," *World Economic Forum*, July 10, 2015.
45. Jake Spring, "Residents of China's Capital Getting Out of Town to Escape Smog," *Reuters*, December 21, 2016. www.reuters.com/article/us-china-pollution-travel/residents-of-chinas-capital-getting-out-of-town-to-escape-smog-idUSKBN14A15H.

46. Yang Xie, et al., "Economic Impacts from PM2.5 Pollution-Related Health Effects in China: A Provincial-Level Analysis," *Environmental Science & Technology* 50, no. 9 (May, 2016): 4836–4843.
47. World Bank, *Cost of Pollution in China*; Yang Xia, et al., "Assessment of Socioeconomic Costs to China's Air Pollution."
48. Yan Jiuyyuan, "Benci dahui dui zhongguo yingxiang zuida de defang" ("The Area Where this Congress Has the Most Impact on China"), October 25, 2017. https://mp.weixin.qq.com/s/BLHWqQLZDTOg7RAas4C_sQ.
49. "Woguo mianlin dieluo zhongdeng shouru xianjing fengxian" ("Our Country Faces Risks of Falling into Middle-income Trap"), *Caijing*, April 14, 2014. http://finance.sina.com.cn/review/hgds/20140414/132718790328.shtml.
50. "Ha Jiming zuixin yanjiang" ("The Latest Speech of Ha Jiming"), May 19, 2016. http://news.hexun.com/2016-05-19/183941955.html.
51. Tom Orlik, "Unrest on Rise as Economy Booms," *Wall Street Journal*, September 26, 2011. http://online.wsj.com/article/SB10001424053111903703604576587070600504108.html.
52. Ben Westcott and Natalie Croker, "70 Years of the People's Republic of China in Five Graphics," *CNN*, October 1, 2019. www.cnn.com/2019/10/01/asia/china-70-anniversary-five-graphs-intl-hnk/index.html.
53. Dingxin Zhao, "The Mandate of Heaven and Performance Legitimation in Historical and Contemporary China" *American Behavioral Scientist* 53, no. 3 (October, 2009): 416–433.
54. Dingxin Zhao, *Hefaxing de zhengzhi: Dangdai zhongguo de guojia yu shehui guanxi (Politics of Legitimacy: The State-Society Relations in Contemporary China*) (Taipei: NTU Press, 2017).
55. Xi Jinping, Remarks on *Beijing wanbao* (*Beijing Evening News*), September 28, 2014.
56. Wu Xianhua, *Wumai wuran paifang de yingxiang jiqi guankong youhua* (*The Impact of Smog Pollution Emission and Its Optimal Control*).
57. Albert O. Hirschman, *Exit, Voice, and Loyalty* (Cambridge, MA: Harvard University Press, 1970).
58. Zhou Jing and Chen Ning (eds.), "Dont Pay for the 870 Smog Economy," *Renmin wang* (People's Network), October 28, 2014. http://yuqing.people.com.cn/n/2014/1028/c354318-25920566.html
59. Thomas Johnson, Arthur P. J. Mol, Lei Zhang, et al., "Living under the Dome: Individual Strategies against Air Pollution in Beijing," *Habitat International* 59 (January, 2017): 110–117.
60. Andrew Szasz, *Shopping Our Way to Safety: How We Changed from Protecting the Environment to Protecting Ourselves* (Minneapolis, MN: University of Minnesota Press, 2007).
61. Liu Yanhua, "Guowuyuan canshi Liu Yanhua: Weile GDP xisheng yidairen jinakang shizai bu kequ" ("It is Unwise to Sacrifice One Generation's Health as Cost for GDP Growth"), *Fenghuang caijing*, January 17, 2017. http://finance.ifeng.com/a/20170117/15150104_0.shtml.

62. Wei Gu, "Almost Half of Wealthy Chinese Want to Leave, Study Shows," *Wall Street Journal*, September 15, 2014. http://blogs.wsj.com/chinarealtime/2014/09/15/almost-half-of-wealthy-chinese-want-to-leave/).
63. "Wealthy Chinese are Fleeing the Country Like Mad," A ChinaFile Infographic Translation, February 3, 2015. www.chinafile.com/multimedia/infographics/wealthy-chinese-are-fleeing-country-mad.
64. Yu Qin and Hongjia Zhu, "Run Away? Air Pollution and Emigration Interests in China," *Journal of Population Economics* (2015): 1–32.
65. Liu Qin, "Citizens leaving China Should Pay Environmental Levy," *chinadialogue*, October 2, 2014. www.chinadialogue.net/blog/6715-Citizens-leaving-China-should-pay-environmental-levy/en.
66. Wang Huiyao, *Guoji rencai lanpi shu: Zhongguo guoji yimin baogao 2014* (*Annual Report on Chinese International Immigration, 2014*) (Beijing: Shehui kexue wenxian chubanshe, 2014).
67. Pinkle Xu, *Immigration and the Chinese HNWI 2017,* Hurun Research, July 15, 2017. www.hurun.net/EN/Article/Details?num=51636DE2A1F4.
68. Elisabeth Lilja, "A New Ecology of Civil Society II," *Journal of Civil Society* 11, no. 4 (December, 2015): 402–407.
69. Feng Jianmin, "Pollution a Reason to Emigrate," *Shanghai Daily*, January 23, 2014, A2. www.shanghaidaily.com/Business/economy/Pollution-a-reason-to-emigrate/shdaily.shtml.
70. Wang Huiyao, *Guoji rencai lanpi shu: Zhongguo guoji yimin baogao 2014.*
71. Frank Shyong, "Visa Program for Wealthy Investors Maxed Out by Chinese Demand," *Los Angeles Times*, August 29, 2014. www.latimes.com/local/la-me-0830-chinese-visas-20140830-story.html.
72. Lin Hang, "Huanjing baohu: chaoyue huanbao zhuyi" ("Environmental Protection: Transcending Environmentalism"), *Phoenix Weekly*, January 5, 2013. www.ifengweekly.com/detil.php?id=174.
73. Zheng Hangsheng, *Jiansuo daijia yu zengcu jinbu: shehuixue jiqi shenceng linian* (*Reducing the Cost and Promoting Progress: Sociology and its Deep Philosophy*) (Beijing: Beijing shifan daxue chubanshe, 2007), 198.
74. John Kennedy, "Environmental Protests in China on Dramatic Rise, Expert Says," *South China Morning Post*, October 29, 2012. www.scmp.com/news/china/article/1072407/environmental-protests-china-rise-expert-says.
75. Elizabeth C. Economy, "The Great Leap Backward? The Costs of China's Environmental Crisis," *Foreign Affairs* 86, no. 5 (September/October, 2007): 38–59.
76. Feng Jie and Wang Tao, "Kaichuang: qiujie huanjing qunti xing shijian" ("Opening the Window: Asking for Solutions to the Environmental Mass Incidents), *Nanfang zhoumo* (*Southern Weekend*), November 29, 2012. www.infzm.com/content/83316.
77. Thomas Lennon and Ruby Yang, "The Warriors of Qiugang: A Chinese Village Fights Back," *Yale Environment* 360, January 10, 2011. https://e360.yale.edu/features/the_warriors_of_qiugang_a_chinese_village_fights_back.
78. Wang Tao, Liang Yuejing and Yang Guoyao, "Zhongguo huanjing bingren" ("Environmental Sick Man of China"), *Nanfang zhoumo* (*Southern Weekend*),

November 27, 2014. https://chinadigitaltimes.net/chinese/2014/11/%E5%8D%97%E6%96%B9%E5%91%A8%E6%9C%AB-%E4%B8%AD%E5%9B%BD%E7%8E%AF%E5%A2%83%E7%97%85%E4%BA%BA/.

79. Guillaume Pitron, "China's army of green activists," *Gulf News*, August 11, 2017. https://gulfnews.com/culture/environment/china-s-army-of-green-activists-1.2071769.

80. Yan Yan and Liu Zuyun, "Fengxian shehui lilun fanshixia zhongguo huanjing chongtu wenti jiqi xietong zhili" ("China's Environmental Conflict Issue and its Synergistic Management under Risk Society Theoretical Paradigm"), *Nanjing shida xuebao* (*Journal of Nanjing Normal University*), social science edition, no. 3 (May, 2014): 35.

81. Daniel K. Gardner, *Environmental Pollution in China (What Everyone Needs To Know®)* (New York, NY: Oxford University Press, 2018), 153.

82. Liu Jianqiang, "China's New 'Middle Class' Environmental Protests," *chinadialogue*, February 1, 2013. www.chinadialogue.net/article/show/single/ch/5561-China-s-new-middle-class-environmental-protests.

83. The two types of mass incidents were first raised in Yu Jianrong, "Canyu de kunjing: 2012 nian de shehui chongtu" (Dilemma of participation: Social conflict in 2012), *Nanfeng chuang* (South window), no. 26 (2012): 60–61. For further discussion, see Qin Bing-yu, "Zhongguo shi shengtai zhengzhi: jiyu jinnian lai huanjing quntixing shijian de fenxi" ("Chinese Ecological Politics: Based on an Analysis of The Environmental Group Events in Recent Years"), *Dongbei daxue xuebao* (*Journal of Northeastern University*), social sciences edition, 17, no.5 (September, 2015).

84. Central Party School professor, Communication with author, January 4, 2017.

85. This array of incidents is the result of my own collection based on both Chinese and foreign media reports.

86. Feng and Wang, "Kaichuang: qiujie huanjing qunti xing shijian" ("Opening the Window: Asking for Solutions to the Environmental Mass Incidents").

87. Howard W. French, "Anger in China Rises Over Threat to Environment," *New York Times*, July 20, 2005. www.nytimes.com/learning/teachers/featured_articles/20050720wednesday.html.

88. Tang Hao, "China's 'Nimby' Protests Sign of Unequal Society," *chinadialogue*, May 29, 2013. www.chinadialogue.net/article/show/single/en/6051-China-s-nimby-protests-sign-of-unequal-society.

89. Qin, "Zhongguo shi shengtai zhengzhi," p. 498.

90. Benjamin Van Rooij, "The People vs. Pollution: Understanding Citizen Action against Pollution in China" *The Journal of Contemporary China* 19, no. 63 (January, 2010): 55–79.

91. John Ruwitch, "Angry Chinese Occupy Govt Office, Smash Computers in Environment Protest," *Reuters*, July 28, 2012. www.reuters.com/article/china-environment-protest-idUSL4E8IR24720120728; Wu Liying, "Yakua lixing de zuihou yi gen daocao" ("The Straw that Breaks the Reason's Back"), *Lianhe zaobao*, August 1, 2012. www.zaobao.com.sg/forum/zaodian/shang-hai-tan/story20120801-57134.

92. Austin Ramzy, *"Environmental Protest Blocks Wastewater Pipeline Near Shanghai," Time*, July 28, 2012. http://world.time.com/2012/07/28/environmental-protest-blocks-wastewater-pipeline-near-shanghai/.
93. Tang Hao, "China's 'Nimby' Movement Sign of Unequal Society," *chinadialogue*, May 29, 2013. www.chinadialogue.net/article/show/single/en/6051-China-s-nimby-protests-sign-of-un.
94. Xiaoling Zhang and Gareth Shaw, "New Media, Emerging Middle Class and Environmental Health Movement in China," in *Urban Mobilizations and New Media in Contemporary China* edited by Lisheng Dong, Hanspeter Kriesi, and Daniel Kübler, 101–116 (Farnham, UK: Ashgate, 2015).
95. Feng Yongfeng, interview with author, Beijing, July 1, 2015.
96. Qiao Long, "China Aims for Near-Total Surveillance, Including in People's Homes," *Radio Free Asia*, March 30, 2018. www.rfa.org/english/news/china/surveillance-03302018111415.html; Associated Press, "China's Sharp Eyes Surveillance System Puts the Security Focus on Public Shaming," *South China Morning Post*, October 30, 2018. www.scmp.com/news/china/politics/article/2170834/chinas-sharp-eyes-surveillance-system-puts-security-focus-public.
97. Wang Jingtao (ed.), "PX xiangmu kuitui zhong huhuan jianshou dian de chuxian" ("PX Project: Calling for a Reference Point to Cling to Amidst Breakdown"), *Huanqiu shibao* (*Global Times*), March 31, 2014. http://opinion.huanqiu.com/editorial/2014-04/4943603.html.
98. Professor GQ, WeChat conversation with author, January 4, 2017.
99. Liu Yang (ed.), "Xi Jinping chuxi quanguo shengtai huanjing baohu dahui bing fabiao zhongyao jianghua" ("Xi Jinping Participated in National Ecological Environment Protection Conference and Made Important Speeches"), *Xinhua*, May 19, 2018. www.gov.cn/xinwen/2018-05/19/content_5292116.htm.
100. MEP, "Guojia huanjing baohu shisanwu huanjing yu jiankang gongzuo guihua" ("Environment and Health Work Plan, National Environmental Protection Thirteenth Five Year Plan"), February 2017.
101. David Stanway and James Regan, "Pollution the Big Barrier to Freer Trade in Rare Earths," *Reuters*, March 19, 2012. www.reuters.com/article/us-china-rareearth/pollution-the-big-barrier-to-freer-trade-in-rare-earths-idUSBRE82I08I20120319; Jonathan Kaiman, "Rare Earth Mining in China: The Bleak Social and Environmental Costs," *The Guardian*, March 20, 2014. www.theguardian.com/sustainable-business/rare-earth-mining-china-social-environmental-costs.
102. Jintai Lin, Da Pan, Steven J. Davis, et al., "China's International Trade and Air Pollution in the United States," *Proceedings of the National Academy of Sciences*, 111 no. 5 (February, 2014): 1736–1741.
103. Xujia Jiang, Qiang Zhang, Hongyan Zhao, et al., "Revealing the Hidden Health Costs Embodied in Chinese Exports," *Environmental Science Technology* 49 (March, 2015): 4381–4388; Qiang Zhang, Xujia Jiang, San Tong, et al., "Transboundary Health Impacts of Transported Global Air Pollution and International Trade." *Nature*, 543 (March, 2017): 705–709.

104. Dabo Guan, Xin Su, Qiang Zhang, et al., "The Socioeconomic Drivers of China's Primary PM2:5 Emissions," *Environmental Research Letters* 9, no. 2 (February, 2014): 9.
105. Bryan N. Duncan, "A Space-based, High-resolution View of Notable Changes in Urban NOx Pollution Around the World (2005–2014)," *Journal of Geographical Research: Atmospheres* (January, 2016): 987. https://agupubs.onlinelibrary.wiley.com/doi/epdf/10.1002/2015JD024121.
106. Kim Jeong-su, "NASA and NIER Study Finds that 48% of Particulate Matter Comes from Outside S. Korea," *Hankyoreh*, July 20, 2017. www.hani.co.kr/arti/english_edition/e_international/803654.html.
107. Doyle Rice, "Air Pollution in Asia is Wafting into the USA, Increasing Smog in West," *USA Today*, March 2, 2017. www.usatoday.com/story/weather/2017/03/02/air-pollution-asia-wafting-into-usa-increasing-smog-west/98647354/.
108. Lin, Pan, Davis, et al., "China's International Trade and Air Pollution in the United States."
109. Michaeleen Doucleff, "Why China's Pollution Could Be behind Our Cold, Snowy Winters," NPR, March 8, 2015. www.npr.org/sections/goatsandsoda/2015/03/08/391056439/why-chinas-pollution-could-be-behind-our-cold-snowy-winters.
110. Eri Saikawa, Vaishali Naik, Larry W.Horowitz, et al., "Present and Potential Future Contributions of Sulfate, Black and Organic Carbon Aerosols from China to Global Air Quality, Premature Mortality and Radiative Forcing," *Atmospheric Environment* 43, no. 17 (June, 2009): 2814–2822.
111. University of East Anglia, "Study Reveals Amount of Premature Deaths Linked to International Trade," Phys.org, March 29, 2017. https://phys.org/news/2017-03-reveals-amount-premature-deaths-linked.html.
112. Professor Judith Shapiro, interview with author, Washington, DC, October 11, 2017.
113. Martin Fackler, "Insecticide-Tainted Dumplings from China Sicken 175 in Japan," *New York Times*, February 2, 2008. www.nytimes.com/2008/02/02/world/asia/02japan.html; Lenore Taylor, "Chinese Tinned Peaches High in Lead, Says MP Battling for SPC Ardmona," *The Guardian*, January 23, 2014. www.theguardian.com/business/2014/jan/24/spc-ardmona-tests-lead-chinese-tinned-peaches.
114. Yanzhong Huang, "Are China's Chickens Contaminating America's Plates?" *Foreign Policy*, November 16, 2017. https://foreignpolicy.com/2017/11/16/are-chinas-chickens-contaminating-americas-plates/.
115. Hu Tao, interview with author, Washington, DC, October 11, 2017.
116. Hu Tao, "The China-US Trade Deal: Environmental Benefits and Costs," *chinadialogue*, May 17, 2017. www.chinadialogue.net/blog/9793-The-China-US-trade-deal-environmental-benefits-and-costs/en.
117. Lydia McMullen-Laird, Xiaofan Zhao, Mengjie Gong, et al., "Air Pollution Governance as a Driver of Recent Climate Policies in China," *Carbon & Climate Law Review* 3, no. 9 (March, 2015): 243–255.

118. Kong Lingyu, "Zhongguo qicao wancheng qihou bianhua yingdui fa caoan" ("China Completed the Draft for Climate Change Handling Law"), *Caixin wang*, July 23, 2014. http://china.caixin.com/2014-07-23/100707692.html.
119. CCTV, "Shouzhu lushui qingshan" ("Protect the Green Water and Green Mountains"), http://tv.cntv.cn/video/C16624/dc2c371a868949c991c38a04385094fe.
120. Tom Philipps, "China's Xi Jinping Says Paris Climate Deal Must Not be Allowed to Fail," *The Guardian*, January 18, 2017. www.theguardian.com/world/2017/jan/19/chinas-xi-jinping-says-world-must-implement-paris-climate-deal.
121. Jay Lemery and Paul Auerbach, *Environmedics: The Impact of Climate Change on Human Health* (New York, NY: Rowman & Littlefield, 2017), 56.
122. Zheng Jinran, "China, South Korea and Japan Vow to Jointly Tackle Air Pollution," *China Daily*, August 25, 2017. www.chinadaily.com.cn/china/2017-08/25/content_31111617.htm.
123. Tom Holland, "Breathe Easy, Neither Trump nor Critics of Paris Pullout Will Affect Climate," *South China Morning Post*, June 5, 2017. www.scmp.com/week-asia/opinion/article/2096705/breathe-easy-neither-trump-nor-critics-paris-pullout-will-affect.
124. Interview with Professor Chih-yu Shih, Guangzhou, December 10, 2019; Jacob Mardell, "The 'Community of Common Destiny' in Xi Jinping's New Era," *The Diplomat*, October 25, 2017. https://thediplomat.com/2017/10/the-community-of-common-destiny-in-xi-jinpings-new-era/.
125. Joseph Nye, "The Kindleberger Trap," *Project Syndicate*, January 9, 2017. www.project-syndicate.org/commentary/trump-china-kindleberger-trap-by-joseph-s–nye-2017-01?barrier=accessreg.
126. Edward Wong, "Nearly 14,000 Companies in China Violate Pollution Rules," *The New York Times*, June 13, 2017.
127. Professor Judith Shapiro, interview with author, Washington, DC, October 11, 2017.
128. Earl Carr and Yan Chen, "Is China Serious about Curbing Pollution along the Belt and Road?" *South China Morning Post*, December 11, 2017.
129. Elena F. Tracy, Evgeny Shvarts, Eugene Simonov, et al., "China's New Eurasian Ambitions: The Environmental Risks of the Silk Road Economic Belt," *Eurasian Geography and Economics* 58, no. 1 (August, 2017): 56–88.
130. Jeff Kearns, Hannah Dormido and Alyssa McDonald, "China's War on Pollution Will Change the World," *Bloomberg*, March 9, 2018. www.bloomberg.com/graphics/2018-china-pollution/.
131. Kearns, Dormido and McDonald, "China's War on Pollution Will Change the World."
132. Thomas Biesheuvel, "China Sends One of the West's Most Critical Materials Soaring," Bloomberg, September 10, 2017. www.bloomberg.com/news/articles/2017-09-10/china-sends-one-of-the-west-s-most-critical-materials-soaring.
133. Cassandra Profita, "Chinese Ban On Waste Imports Puts Northwest Recycling In Limbo," *OPB*, October 2, 2017. www.opb.org/news/article/chinese-ban-waste-north west-recycling-limbo/; Pak Yiu, "China ban on waste imports leads to piles of paper abroad, surging prices in China," *Reuters*, September 28, 2017. www.reuters.com/

article/us-china-hongkong-paperrecycling/china-ban-on-waste-imports-leads-to-piles-of-paper-abroad-surging-prices-in-china-idUSKCN1C30GR.

134. Tom Baxter and Liu Hua, "24 Reasons Why China's Ban on Foreign Trash is a Wake-up Call for Global Waste Exporters," *South China Morning Post*, December 31, 2017. www.scmp.com/comment/insight-opinion/article/2126098/24-reasons-why-chinas-ban-foreign-trash-wake-call-global.
135. Global Times, "Yong PM2.5 guanxin zhongguo huanbao de meiguoren haishi luchu le zhenmianmu" ("The Americans Who Use PM2.5 to Care about China's Environmental Protection Have Showed their True Colors"), *Sina*, September 13, 2017. https://news.sina.cn/gj/2017-09-13/detail-ifykuftz6697294.d.html.
136. Taiyuan Environmental Protection Bureau official, interview with author, Taiyuan, December 14, 2015.
137. Zhou Xiaoping, "Buyao yinwei wumai er zangsong le zhongguo de gongye" ("Don't Sacrifice China's Industry Because of Smog"), *Sina*, December 14, 2016. http://cj.sina.com.cn/article/detail/2949462582/124415?cre=tianyi&mod=wpage&loc=1&r=9&doct=0&rfunc=0&tj=none&s=0&tr=9.
138. Zhou Xiaoping, "Zhimai de weiyi fangshi zhongyu zhaodao le" ("The Only Solution to the Smog Problem Has Finally Been Found"), *Sina Blog*, December 21, 2016. http://blog.sina.com.cn/s/blog_48a082b70102x60j.html?tj=1.
139. Qian Weihong and Ding Zhongli, "Face to Face." Interview by CCTV, March 21, 2010. http://tv.cntv.cn/video/C10359/ffd0945e07184006a826c5bcf1eb9c35.
140. Liyan Qi and Te-Ping Chen, "Chinese Scientist Blasts Trump's Climate-Change Talk," *Wall Street Journal*, January 27, 2017. https://blogs.wsj.com/chinarealtime/2017/01/27/0127cpollute/.
141. Josh Horwitz, "A Chinese Student's Commencement Speech Praising 'Fresh Air' and Democracy is Riling China's Internet," *Quartz*, May 23, 2017. https://qz.com/989454/a-chinese-students-commencement-speech-at-the-university-of-maryland-praising-fresh-air-and-democracy-is-riling-chinas-internet/.
142. "Malilan daxue shi fangong fanhua jidi" ("University of Maryland is the Anti-communist and Anti-China Base"), *china.com*, May 25, 2017. http://military.china.com/jctp/11172988/20170525/30565795_all.html.
143. Professor XL, Communication with author, May 24, 2016.
144. Yinan He, "Domestic Troubles, National Identity Discourse, and China's Attitude towards the West, 2003–2012," *Nations and Nationalism* 24, no. 3 (July, 2018): 741–766.

EVOLVING ENVIRONMENTAL HEALTH POLICY

1. Ma Jun, Interview with author, Beijing, July 1, 2015.
2. Cui Huqin, "Chai Jing's Under the Dome: A multimedia documentary in the digital age," *Journal of Chinese Cinemas* 11, no. 1 (January, 2017): 30–45.
3. Local environmental activists, Interview with author, Guangzhou, July 11, 2015.
4. Yu Yong, Li Yan, "Qu Geping: Zhou zongli bi chu le xinzhongguo di yi dai huanbao ren" ("Qu Geping: Premier Zhou 'Forced Out' a New Generation of People Dedicated to

Environmental Protection in New China"), *Sina History*, January 8, 2015. http://history.sina.com.cn/bk/ds/2015-01-08/1127114596.shtml.

5. Liu Xuesong, "Mao Zedong yu xinzhongguo yiliao weisheng gongzuo" ("Mao Zedong and Medical and Healthcare Work"), *Renmin wang*, May 9, 2016. http://dangshi.people.com.cn/n1/2016/0509/c85037-28333912-3.html.
6. Yanzhong Huang, *Governing Health in Contemporary China* (New York: Routledge, 2013), 32–33.
7. "Liang Sicheng yu Mao Zedong de lao Beijing cheng cunfei zhi zheng" ("The Debate between Liang Sicheng and Mao Zedong over the Reserving and Abolishing over the Old Beijing Downtown"), *Fenghuang wang*, January 19, 2009. http://phtv.ifeng.com/program/ffdgm/200901/0119_2309_976064_5.shtml.
8. Elizabeth Economy, *The River Runs Black: The Environmental Challenge to China's Future* (Ithaca, NY: Cornell University Press, 2004), 51–55.
9. Yang Wenli, "Zhou Enlai yu zhongguo huanjing baohu gongzuo de qibu" ("Zhou Enlai and the Start of China's Environmental Protection Work"). *Dangdai zhongguo shi yanjiu* (*Contemporary China History Studies*), March 2008.
10. Yu and Li, "Qu Geping: Zhou zongli bi chu le xinzhongguo di yi dai huanbao ren" ("Qu Geping: Premier Zhou 'Forced Out' a New Generation of People Dedicated to Environmental Protection in New China").
11. Liu Dong, "Zhou Enlai guanyu huanjing baohu de lunshu yu shijian" ("Zhou Enlai and the Theory and Practice on Environmental Protection"), *Beijing Academy of Social Sciences*, 2006. http://cpc.people.com.cn/GB/69112/75843/75873/5168346.html.
12. Yu and Li, "Qu Geping: Zhou zongli bi chu le xinzhongguo di yi dai huanbao ren" ("Qu Geping: Premier Zhou 'Forced Out' a New Generation of People Dedicated to Environmental Protection in New China").
13. Beginning in 1968, the Ministry of Health, like many other central ministries during the Cultural Revolution, saw army representatives assume control of its day-to-day operations. The military control did not end until 1973.
14. Yu and Li, "Qu Geping: Zhou zongli bi chu le xinzhongguo di yi dai huanbao ren" ("Qu Geping: Premier Zhou 'Forced Out' a New Generation of People Dedicated to Environmental Protection in New China").
15. Wang Tao, "Bierang Beijing chengwei lundun nayang de wudu" (Don't Let Beijing Become the City of Fog Like London), March 29, 2013. www.cqvip.com/qk/84412x/201701/83686671504849554849485049.html.
16. Zhen Zhong, "The Dynamic Evolution of China's Environment Policy," School of Agricultural Economics and Rural Development, Renmin University of China, September 23, 2015. http://ap.fftc.agnet.org/ap_db.php?id=506&print=1.
17. Qu Geping, *Mengxiang yu qidai: Zhongguo huanjing baohu de guoqu yu weilai* (*Dreams and Expectations: The Past and Future of China's Environmental Protection*), (Beijing: Zhongguo huanjing kexue chubanshe, 2000), 40–41. On policy primeval soup, see John Kingdon, *Agendas, Alternatives, and Public Policies*, Second Edition (Boston, MA: Little, Brown, 1984), 143.

18. Wu Dui, a Guangzhou-based environmental scientist, started to collect PM2.5 samples as early as 1989. Local environmental activist, Interview with author, Guangzhou, July 11, 2015.
19. Wang, "Bierang Beijing chengwei lundun nayang de wudu" ("Don't Let Beijing Become the City of Fog Like London").
20. Jing Fang and Gerry Bloom, "China's Rural Health System and Environment-Related Health Risks." *Journal of Contemporary China* 19, no. 63 (January, 2010): 33.
21. Qing Lan, Robert Sedwick Chapman, Xingzhou He, et al., "Household Stove Improvement and Risk of Lung Cancer in Xuanwei, China," *Journal of the National Cancer Institute* 94 (June, 2002): 826–835.
22. Wang Tao, "Wumai jiankang yanjiu li de wumai sanshi nanli" ("After Thirty Years the Research on Haze and Health Remains Fuzzy"), *Nanfang zhou mo* (*Southern Weekend*), March 29, 2013. www.infzm.com/content/89147.
23. Former Mayor of Yangzhong, Interview with author, August 2015.
24. National Development and Reform Commission, Ministry of Water Resources, Ministry of Health, "Quanguo nongcun yinshui anquan gongcheng shiyiwu guihua qianyan" ("Preface, The Eleventh Five Year Plan for National Rural Drinking Water Safety Project"), 2008. www.ndrc.gov.cn/fzgggz/fzgh/ghwb/gjjgh/200804/P020150630514326308864.pdf.
25. Xi Penghui, Liang Ruobing, "Chengshi kongqi zhiliang yu huanjing yimin" ("Urban Air Quality and Environmental Migration)," *Jingji kexue* (*Economic Science*), no. 4, (2015): 30–43.
26. Ruth Levine, et al., "CASE 15: Preventing Iodine Deficiency Disease in China," Center for Global Development. www.cgdev.org/page/case-15-preventing-iodine-deficiency-disease-china.
27. Economy, *The River Runs Black*, chapter 4.
28. *Beijing wanbao* (*Beijing Evening News*), March 9, 1999. www.sohu.com/a/122658925_573336.
29. Luo Dawei, "PM2.5 de zhongguo lu" (PM2.5's China path), *Wangyi zhenhua (Netease Truth)*, February 22, 2014. http://luodawei.siyuefeng.com/article/8823.
30. Li Min-ming, Jia Cao, Jian Xu, et al., "The National Trend of Blood Lead Levels among Chinese Children Aged 0–18 Years Old, 1990–2012," *Environment International* 71 (October, 2014): 109–117.
31. Dabo Guan, Xin Su, Qiang Zhang, et al., "The Socioeconomic Drivers of China's Primary PM2.5 Emissions," *Environmental Research Letters* 9, no. 2 (February, 2014). https://iopscience.iop.org/article/10.1088/1748-9326/9/2/024010/meta.
32. David Barboza, "China Reportedly Urged Omitting Pollution-Death Estimates," *New York Times*, July 5, 2007. www.nytimes.com/2007/07/05/world/asia/05china.html.
33. Wikileaks, "Waiting to Inhale: The State of Air Pollution in South China (Part I of II)," August 16, 2006. Canonical ID: 06GUANGZHOU27482_a https://wikileaks.org/plusd/cables/06GUANGZHOU27482_a.html.
34. Xie Haitao, Liu Hongqiao, "Huaihe aizheng" ("Huaihe Cancer"), *Caixin Xinshiji* (*Caixin New Century*). www.letscorp.net/archives/55533.

35. Professor Yang Gonghuan, Interview with author, Beijing, July 2, 2015.
36. Jere Longman, "Olympics; Beijing Wins Bid for 2008 Olympic Games," *The New York Times*, July 14, 2001.
37. Jin Dapeng, Arne Ljungqvist and Hans Troedsson (eds.), *The Health Legacy of the 2008 Beijing Olympic Games: Successes and Recommendations* (Geneva: World Health Organization, 2010), 108.
38. Gregory Carmichael, Soon-Chang Yoon, Cy Jones, et al., Independent Environmental Assessment: *Beijing 2008 Olympic Games.* United Nations Environment Programme, February 2009. http://hdl.handle.net/20.500.11822/7562.
39. Michael Lelyveld, "China Warned on Olympic Pollution," *Radio Free Asia*, November 2, 2007. www.rfa.org/english/features/lelyveld/china_lelyveld-20071102.html.
40. Wang Zhi, "Zhijing de bujin shi sudu" ("It is Not Just Speed That is to be Saluted To"), *Changjiang shangbao* (*Changjiang Times*), March 22, 2015. www.changjiangtimes.com/2015/03/498673.html; Long Jing, "Huimai longzhao xia kongqi ye tegong le" ("Air under Smog is also Under 'Special Supply'"), *Shidai zhoubao* (*The Time Weekly*), November 14, 2011. http://news.ifeng.com/opinion/gundong/detail_2011_11/14/10648029_0.shtml.
41. Andrew Jacobs, "The Privileges of China's Elite Include Purified Air," *New York Times*, November 4, 2011. www.nytimes.com/2011/11/05/world/asia/the-privileges-of-chinas-elite-include-purified-air.html.
42. Austin Ramzy, "Twittering Bad Air Particles in Beijing," *Time*, June 19, 2009. http://content.time.com/time/printout/0,8816,1905736,00.html.
43. Liu Zhengxu, "Zhong Nanshan: Shuju xianshi wushi sui yishang Guangzhou ren feizang cheng heise" ("Zhong Nanshan: Data Shows that the Lungs of Guangzhou Residents Aged above 50 Turned Black"), *Xin kuai bao*, June 13, 2008. http://news.sina.com.cn/c/2008-06-13/004814008410s.shtml.
44. Peony Lui, "Pollution is One Reason Guangzhou People's Lungs are Turning Black, Warns Expert," *South China Morning Post*, March 7, 2013. www.scmp.com/news/china/article/1185398/pollution-one-reason-guangzhou-peoples-lungs-are-turning-black-warns
45. "Meiguo yundongyuan dai kouzao wuru le shui?" ("Who Was Humiliated by American Athletes Wearing Masks?"), *Renmin wang*, August 7, 2008. www.people.com.cn/GB/32306/33232/7626530.html.
46. Deborah Charles, "American Cyclists Apologize for Masks." *Reuters*, August 6, 2008. www.reuters.com/article/us-olympics-masks-apology/american-cyclists-apologize-for-masks-idUSPEK33436220080806.
47. "Beijing huanbaoju huiying guowai yundongyuan dai kouzhao canjia aoyunhui" (Beijing EPB Responded to News that Foreign Athletes Were Planning to Wear Masks to Participate in the Olympic Games). July 25, 2008. http://news.sina.com.cn/c/2008-07-25/154716000354.shtml.
48. US Embassy officials, Interview with author, Beijing, July 6, 2015.
49. US Embassy officials, Interview with author, Beijing, July 6, 2015.

50. Wenzhong Yan Yu, "PM2.5 jin zhongguo ji guanfang wunianqian zai shige chengshi shidian jiance" ("The Story of PM2.5 Entering China: The Government Piloted Monitoring in 10 Cities Ten Years Ago"), *Zhongguo zhoukan* (*China Weekly*), August 21, 2012. http://news.sina.com.cn/c/sd/2012-08-21/114925007148.shtml.
51. Diplomatic cable 09BEIJING1945, available at http://wikileaks.org/cable/2009/07/09BEIJING1945.html.
52. Didi Kirsten Tatlow, "China Has Made Strides in Addressing Air Pollution, Environmentalist Says," *New York Times*, December 19, 2016.
53. Ramzy, "Twittering Bad Air Particles in Beijing."
54. BBC, "Zhongguo qing zhuhua lingshiguan tingfa kongqi zhiliang shuju" ("China Asked Foreign Consulates in China to Stop Issuing Air Quality Data"), *BBC News*, June 5, 2012. www.bbc.com/zhongwen/simp/chinese_news/2012/06/120605_china_air_quality
55. US Embassy officials, Interview with author, Beijing, July 6, 2015.
56. Diplomatic cable 09BEIJING1945, available at http://wikileaks.org/cable/2009/07/09BEIJING1945.html.
57. James Fallows, "In China, 'Time Is Not Ripe' for Honest Air Pollution Readings," *The Atlantic*, November 11, 2011. www.theatlantic.com/international/archive/2011/11/in-china-time-is-not-ripe-for-honest-air-pollution-readings/247817/.
58. Du Juan, "PM2.5 na pingjia biaozhun hou zhongguo bacheng chengshi kongqi bu dabiao" ("80% of Chinese Cities Cannot Meet the Standards for Air Quality after Introducing PM2.5 into the Evaluation System"). *Guangzhou ribao*, November 16, 2011. http://finance.sina.com.cn/china/dfjj/20111116/093510825206.shtml.
59. Ai Li Xin, "Beijing de kongqi wuran cheng zhengzhi wenti" ("Is Beijing's Air Pollution Becoming a Political Issue?"), *DW News*, October 1, 2012. https://p.dw.com/p/13NPW
60. Si-ming Lu, "A Case Study of Risk Communication: The Beijing Smog: The Communication Battle between the Public and Government," 2016 International Conference on Education, Management and Applied Social Science (EMASS, 2016). www.dpi-proceedings.com/index.php/dtssehs/article/viewFile/6804/6396.
61. Huang Chong, "Jing qicheng shoufangzhe ganjue kongqi jiance shuju he zhiguan ganshou bufu" ("Nearly 70 Percent of the Respondents Said that the Government Air Monitoring Data Was Not Compatible with What They Actually Felt"), *Zhongguo qingnian bao* (*China Youth Daily*), November 8, 2011, 7.
62. Jonathan Watts, "Twitter Gaffe: US Embassy Announces 'Crazy Bad' Beijing Air Pollution," *The Guardian*, November 19, 2010. www.theguardian.com/environment/blog/2010/nov/19/crazy-bad-beijing-air-pollution.
63. Cui Zheng, "Beijing qingwei wuran de mimi" ("The Secret of Beijing Being 'Lighted Polluted'"), *Caixin wang*, October 24, 2011. http://opinion.caixin.cn/2011-10-24/100316977.html.
64. Jaime A. FlorCruz, "Beijing's New Year surprise: PM 2.5 Readings," *CNN*, January 27, 2012. www.cnn.com/2012/01/27/world/asia/florcruz-china-pollution/index.html.
65. Feng Jie and Lv Zongshu, "Wo wei zuguo ce kongqi" ("I Gauge Air Quality for My Motherland"), *Nanfang zhoumo* (*Southern Weekend*), October 28, 2011. www.infzm.com/content/64281.

66. Yu Meng, "Liu yue qi meiri bobao PM2.5" ("PM2.5 Level Will be Broadcast Daily Beginning in June"), *Dongfang zaobao* (*Oriental Morning Post*), February 28, 2012. https://news.qq.com/a/20120228/000427.htm.
67. Zhang Ke, "Woguo yi jiancheng fazhanzhong guojia zuida huanjing kongqi zhiliang jiance wang" ("Our Country Has Built the Largest Environmental Air Quality Monitoring Network among Developing Countries"), *Diyi caijing (yicai)*, January 31, 2018. www.yicai.com/news/5396870.html.
68. Yu Meng, "Liu yue qi meiri bobao PM2.5" ("PM2.5 Level Will be Broadcast Daily Beginning in June").
69. Dalia Ghanem and Junjie Zhang, "'Effortless Perfection:' Do Chinese Cities Manipulate Air Pollution Data?" *Journal of Environmental Economics and Management* 68, no. 2 (September, 2014): 203–225.
70. Professor Ling Liguo, Interview with author, Shanghai, June 25, 2015.
71. Wang Qian, "China to 'Declare War' on Pollution, Cut Energy Use," *China Daily*, March 14, 2014. www.chinadaily.com.cn/china/2014npcandcppcc/2014-03/14/content_17346330.htm.
72. Edward Wong, "New Details of How Wife of Chinese Politician Thought She Was Poisoned," *New York Times*, October 15, 2012. https://cn.nytimes.com/china/20121015/c15gu/dual/.
73. BBC, "Zhongguo qing zhuhua lingshiguan tingfa kongqi zhiliang shuju" ("China Asked Foreign Consulates in China to Stop Issuing Air Quality Data").
74. Anne Henochowicz, "Netizen Voices: Clearing the Air," *China Digital Times*, June 6, 2012. https://chinadigitaltimes.net/2012/06/netizen-voices-clearing-the-air/.
75. Henochowicz, "Netizen Voices: Clearing the Air."
76. Susan Shirk and Steven Oliver, "China Has No Good Answer to the US Embassy Pollution-Monitoring," *The Atlantic*, Jun 13, 2012. www.theatlantic.com/international/archive/2012/06/china-has-no-good-answer-to-the-us-embassy-pollution-monitoring/258447/1.
77. "Lushui qingshan jiushi jinshan yinshan" ("Clear Water and Green Mountains are as Good as Mountains of Gold and Silver"), *Zhejiang ribao* (*Zhejiang Daily*), October 9, 2017. http://news.12371.cn/2017/10/09/ARTI1507519467134632.shtml.
78. Zhang Yulin, "Zhengjing yitihua kaifa jizhi yu zhongguo nongcun de huanjing chongtuo" ("Development Mechanism of Political-economic Integration and the Environmental Conflicts in China's Countryside"), Department of Sociology, Nanjing University, February 20, 2007. http://ww2.usc.cuhk.edu.hk/PaperCollection/Details.aspx?id=5920.
79. "Lushui qingshan jiushi jinshan yinshan" ("Clear Water and Green Mountains are as Good as Mountains of Gold and Silver").
80. "Lushui qingshan jiushi Jinshan yinshan shi zheyang laide" (This is where "Clear water and green mountains are as good as mountains of gold and silver" is from), *Banyuetan wang*, October 13, 2016. http://news.xinhuanet.com/city/2016-10/13/c_129320831.htm.
81. T.P., "Blackest Day," *The Economist*, January 14, 2013. www.economist.com/analects/2013/01/14/blackest-day.

82. Gwynn Guilford, "China Now Has up to 400 'Cancer Villages,' and the Government Only Just Admitted It," *Quartz*, February 22, 2013. https://qz.com/55928/china-now-has-up-to-400-cancer-villages-and-the-government-only-just-admitted-it/.
83. "2013nian quanguo wumai tianshu yi da 52 nianlai zhi zui" ("The Number of Smog Days in 2013 is the Highest in the Past 52 Years"), *Tianqi*, December 30, 2013. www.tianqi.com/news/21695.html.
84. "Huanbaobu: gebie chengshi wumai chao 200 tian" ("The Number of Smog Days in Some Cities Exceeds 200"), *Renmin wang*, March 15, 2013. http://news.sina.com.cn/c/2013-03-15/153626543243.shtml.
85. Gwynn Guilford, "China's Northeast Hit by Air Pollution So Bad 'You Can't See Your Own Fingers in Front of You,'" *Quartz*, October 21, 2013. https://qz.com/137562/chinas-northeast-hit-by-air-pollution-so-bad-you-cant-see-your-own-fingers-in-front-of-nyou/.
86. WHO, "Outdoor Air Pollution a Leading Environmental Cause of Cancer Deaths," World Health Organization Regional Office for Europe, October 10, 2013. www.euro.who.int/en/health-topics/environment-and-health/urban-health/news/news/2013/10/outdoor-air-pollution-a-leading-environmental-cause-of-cancer-deaths.
87. "Assessment of Haze-related Human Health Risks for Four Chinese Cities During Extreme Haze in January 2013," *National Medical Journal China* 93, no. 34 (September, 2013) https://wenku.baidu.com/view/00e8c1550b1c59eef8c7b4fe.html.
88. Mu Quan and Zhang Shiqiu, "2013 nian yi yue zhongguo damianji wumai shijian zhijie shehui jingji sunshi pinggu" ("An Assessment of Direct Socio-economic Losses Incurred by the Large-scale Smog Incident in China in January 2013"), *Zhongguo huanjing kexue* (*China Environmental Science*), November, 2013.
89. Pan Xiaochuan, Liu Liqun, Zhang Siqi, et al., *Daqi PM2.5 dui zhongguo chengshi gongzhong jiankang xiaoying yanjiu* (*A Study on the Health Effects of Atmospheric PM2.5 Urban China*), Beijing: Kexue chubanshe (January 1, 2016).
90. Jonathan Kaiman, "Chinese Struggle Through 'Airpocalypse' Smog," *The Guardian*, February 16, 2013. www.theguardian.com/world/2013/feb/16/chinese-struggle-through-airpocalypse-smog.
91. BBC, "Zhongguo zhengfu mingdan biaojue huanbao buzhang de piao zuidi" ("Minister of Environmental Protection Received the Lowest Votes in the Chinese Government List"), *BBC*, March 16, 2013. www.bbc.com/zhongwen/simp/china/2013/03/130316_china_npc_government.shtml.
92. Didi Tatlow, "As Cancer Rates Rise in China, Trust Remains Low," *New York Times*, April 17, 2013.
93. The National Environmental Health Action Plan was officially unveiled by the MEP.
94. Kingdon, *Agendas, Alternatives, and Public Policies*.
95. Stephen Chen, "Smog Crisis in China Leads to Increased Research into Effect of Pollution on Fertility," *South China Morning Post*, December 11, 2013. www.scmp.com/news/china/article/1378103/smog-crisis-china-leads-increased-research-effect-pollution-fertility.
96. Wang Qian, "China to 'Declare War' on Pollution, Cut Energy Use," *China Daily*, March 14, 2014. www.chinadaily.com.cn/china/2014npcandcppcc/2014-03/14/content_17346330.htm.

97. Zou Chunxia, "Qi changwei qunian 560 ci pishi huanbao" ("Seven Standing Committee Members Gave 560 Written Instructions in the Past Year"), *Beijing qingnian bao* (*Beijing Youth Daily*), January 16, 2015. http://news.sohu.com/20150116/n407832428.shtml.
98. David Stanway and Kathy Chen, "China's Xi Says Smog His Top Priority during APEC Meet," *Reuters*, November 10, 2014. https://af.reuters.com/article/worldNews/idAFKCN0IV0A120141111.
99. Liu Dong, "Tipping the Scale: Beijing Leads in Obesity Rate," *China Daily*, June 28, 2017. www.chinadaily.com.cn/china/2017-06/28/content_29921273.htm.
100. "Lianheguo huanjingshu zhuren jiedu shijiuda" ("The Executive Director of the United Nations Environment Programme Interprets the 19th Party Congress"), *Zhongguo huanjing bao* (China Environment News), November 23, 2017. www.cenews.com.cn/news/word/201711/t20171123_858941.html.
101. Lan Xue and Jing Zhao, "Truncated Decision Making and Deliberate Implementation: A Time-Based Policy Process Model for Transitional China," *Policy Studies Journal*, 48, no. 2 (May, 2020), 298–326.
102. Daniel M. Fox, *The Convergence of Science and Governance* (California: University of California Press, 2010).
103. Professor LC, Interview with author, Shanghai, March 8, 2017.
104. Qingyue Meng, et al., *Health Policy and Systems Research in China*, UNICEF/UNDP/World Bank/WHO Special Programme for Research and Training in Tropical Diseases (TDR), 2004, vii.
105. Li Xi, "Zhongguo wumai yujing buguan jiankang zhi gu shuzi" ("China Smog Warning: Only Care about Number but Not Health"), *The Other Side*. http://view.163.com/special/reviews/aqi1013.html.
106. Li Wei, "Daibiao jianyi gongke wumai xingcheng jili" ("Representative Suggested to Study the Mechanism of Smog Formation"), *Xin jing bao* (*The Beijing News*), March 10, 2017. www.chinanews.com/gn/2017/03-10/8170161.shtml. Encouraged by government subsidies to increase agricultural output, the use of nitrogen fertilizers surged by about fifty-five times between the 1960s and 2010. Niu Shuping, "China Needs to Cut Use of Chemical Fertilizers: Research." *Reuters*, January 14, 2010. www.reuters.com/article/us-china-agriculture-fertiliser/china-needs-to-cut-use-of-chemical-fertilizers-research-idUSTRE60D20T20100114.
107. China-Britain Business Council, "Zhongguo chutai turang wurang fangzhi xingdong jihua" ("China Unveils Action Plan on Soil Contamination"), *China Go Abroad*, May 31, 2016. www.chinagoabroad.com/zh/article/china-issues-soil-pollution-act-to-complement-existing-air-and-water-policies.
108. Wang Tao, "Wumai jiankang yanjiu li de wumai sanshi nanli" (After Thirty Years the Research on Haze and Health Remains Fuzzy).
109. P Yin, M Brauer, A Cohen, et al., "Long-term Fine Particulate Matter Exposure and Nonaccidental and Cause-specific Mortality in a Large National Cohort of Chinese Men," *Environmental Health Perspective* 125, no. 11 (November, 2017): 117002.
110. Wu Wei, et al., "Beijing shi weijiwei: zheng yanjiu wumai yu feiai fabing guanxi" ("Beijing Municipal Health and Family Planning Commission"), *Xin jing bao* (*The Beijing News*),

January 15, 2017. www.chinanews.com/gn/2017/01-15/8124513.shtml; Wang Can, "Zongli tichu de wumai chengyin yanjiu zhuanxiang gongshi" ("Announcement on the Special Research Program on Smog Formation Proposed by the Premier"), September 4, 2017. www.thepaper.cn/newsDetail_forward_1784243.

111. Wu Jiajia, "Guojia weijiwei huiying wumai dui jiankang yingxiang huati" ("NHFPC Responds to Topics on Smog's Impact on Health"), *Jingji ribao* (*Economic Daily*), January 8, 2017. www.ce.cn/xwzx/gnsz/gdxw/201701/08/t20170108_19455032.shtml.

112. Professor DT, phone conversation with author, January 8, 2018.

113. Wang Tao, "Beida jiaoshou: guojia guangzhu daqi wuran jiankang xiaoying yanjiu wan le dian" ("Beida Professor: The Government Has Been Late in Paying Attention to the Health Impact of Air Pollution"), *Nanfang zhoumo* (*Southern Weekend*), April 2, 2013. http://news.ifeng.com/shendu/nfzm/detail_2013_04/02/23791399_0.shtml.

114. Zhang Lin, "Huanjing jiankang yanjiu: you duoshao qianzhang dai bu" ("Environmental Health Research: How Much Outstanding Debt We Owe"), *Zhongguo kexue bao* (*China Science News*), July 1, 2013. http://paper.sciencenet.cn/htmlnews/2013/7/279442.shtm.

115. Didi Kirsten Tatlow, "China Has Made Strides in Addressing Air Pollution, Environmentalist Says," *New York Times*, December 19, 2016. https://cn.nytimes.com/china/20161219/china-air-pollution-ma-jun/en-us/.

116. Wu Pengfei, et al., "Naqian huanming haishi na ming huanqian?" ("Money for Life or Life for Money?"), China Green Media Research Scholarship Group, November 27, 2017. http://chuansong.me/n/2053227152527.

117. Alice Yang, "Thousands of Polluters in Northern China Fake Emissions Data, Resist Checks," *South China Morning Post*, March 31, 2017. www.scmp.com/news/china/policies-politics/article/2083780/thousands-polluters-northern-china-fake-emissions-data.

118. Zhang Ke, "Jiance buzhun zhuanjia cheng jiashuju fanlan zhi daqi wuran zhili shiqu zhicheng" ("Inaccurate Monitoring: Experts Say the Spread of Fake Data Leads to Lack of Support of Air Pollution Governance"), *Diyi caijing (yicai)*, October 30, 2016. http://m.yicai.com/news/5146142.html.

119. "Environmental Health and China's Rise: Insights from a CFR Workshop," Council on Foreign Relations, January 11, 2017. www.cfr.org/report/environmental-health-and-chinas-rise.

120. Gao Jianghong, "Xuezhe cheng shuizhi qingkuang hen zaogao, dan duozao shi guojia jimi" ("Scholars Say the Water Quality is Very Bad, But How Bad is a State Secret"), *21 shiji jingji baodao* (*Financial Report of the 21st Century*), April 27, 2014. http://news.sina.com.cn/c/2014-04-27/050630018936.shtml.

121. "Zhongguo shouci gongbu turang wuran diaocha jieguo" (China for the First Time Unveils Results of Soil Pollution Survey), *Radio Free Asia*, April 17, 2014.

122. Zhang Hongguang, Zheng Lei, "Wumai yanzhong, ta zhichu le wenti suozai, da'an rang women gandao jue wang" ("The Smog is Serious, and He Pinpoints the Problem,

But the Answer Makes us Feel Desperate"), *Sina*, October 6, 2016. http://blog.sina.com.cn/s/blog_52f526870102x305.html.

123. Angel Hsu and William Miao, "Soil Pollution in China Still a State Secret." *Scientific American*, June 18, 2014. https://blogs.scientificamerican.com/guest-blog/soil-pollution-in-china-still-a-state-secret-infographic/.
124. Professor XJ, interview with author, Beijing, July 2, 2015.
125. Li, "Zhongguo wumai yujing buguan jiankang zhi gu shuzi" ("China Smog Warning: Only Care about Number but Not Health").
126. Zhou Xiaoyuan, "2/3 chengshi kongqi zhiliang bu dabiao" ("Air Quality in 2/3 of the Cities Do Not Meet the Standards"), *Renmin ribao* (*People's Daily*), overseas edition, March 3, 2012. 4. http://paper.people.com.cn/rmrbhwb/html/2012-03/03/content_1015747.htm?div=-1.
127. Zhou Rong, "The Diplomacy of Air Pollution." *chinadialogue*, November 6, 2012. www.chinadialogue.net/article/show/single/en/4971-The-diplomacy-of-air-pollution
128. Li, "Zhongguo wumai yujing buguan jiankang zhi gu shuzi" ("China Smog Warning: Only Care about Number But Not Health").
129. Liu Xiaoping, "Huanbao yu weiji bumen ke gongtong kaizhan diaocha" (Environmental Protection and Health Departments Can Conduct Joint Research), *Fazhi ribao* (*Legal Daily*), March 27, 2017. http://legal.people.com.cn/n1/2017/0327/c42510-29171367.html.
130. Ministry of Environmental Protection, "Guanyu yinfa Guojia huanjing baohu huanjing yu jiankang gongzuo banfa shixing de tongzhi" ("Circular on Printing and Distributing the Trial Approach to Environmental Health in Environmental Protection"), January 25, 2018. www.mee.gov.cn/gkml/hbb/bgt/201801/t20180130_430549_wap.shtml?COLLCC=2332697933&.
131. Bloomberg, "China's Anti-pollution Drive Will Hit Economic Growth, Raise Prices, Economist Says," *South China Morning Post*, October 2, 2017.
132. David Stern, "The Rise and Fall of the Environmental Kuznets Curve," *World Development* 32, no. 8 (March, 2004): 1419–1439.; Richard Carson, "The Environmental Kuznets Curve: Seeking Empirical Regularity and Theoretical Structure," *Review of Environmental Economics and Policy* 4, no. 1 (2010): 3–23.
133. Chang Jiwen, "Fazhan shi jiejue huanjing wenti de guanjian jucuo" ("Development is the Key Measure for Addressing Environmental Problems"), *Renmin ribao* (*People's Daily*), June 6, 2015. http://opinion.people.com.cn/n/2015/0606/c1003-27112174.html
134. Scholar at Jiangsu Environmental Science Academy, Interview with author, Nanjing, June 29, 2015.
135. Lu Yijie and Liu Xing, "Lin Yifu: Jinji zengzhang kuai bushi wumaiduo de juedui yinsu" ("Lin Yifu: Economic Growth is By No Means the Absolute Cause of the Smog Problem"), *Zhongguo qingnianbao* (*China Youth Daily*), March 7, 2015. www.ce.cn/cysc/ny/gdxw/201503/07/t20150307_4751205.shtml. In September 2019, Premier Li Keqiang reaffirmed that economic construction be taken as the central task.
136. Chinese Scholar, WeChat communication with author, August 13, 2017.

137. Tom Holland, "The Risk of the Middle-income Trap Just Increased for China. Here's Why," *South China Morning Post*, October 23, 2017. www.scmp.com/week-asia/opinion/article/2116323/risk-middle-income-trap-just-increased-china-heres-why.

IMPLEMENTING ENVIRONMENTAL HEALTH POLICY

1. Steven Schwankert, "Did Beijing's Mayor Really Say He Would 'Kill Himself' If Air Quality Doesn't Improve?" *The Beijinger*, January 26, 2014. www.thebeijinger.com/blog/2014/01/26/did-beijings-mayor-really-say-he-would-kill-himself-if-air-quality-doesnt-improve.
2. Rao Pie, "Wang Anshun jieshi titou laijian" ("Wang Anshun Explained 'Head on a Platter'"), *Xin jing bao* (*The Beijing News*), January 25, 2015. www.chinanews.com/gn/2015/01–25/7002587.shtml.
3. Qu Geping, Interview by *Qilu Evening News*, March 19, 2014. www.hebeikedun.com/news_detail/newsId=35.html.
4. Keith Crane and Zhimin Mao, *Costs of Selected Policies to Address Air Pollution in China* (Santa Monica, CA: RAND Corporation, 2015). www.rand.org/pubs/research_reports/RR861.html.
5. Carlos Dora, Coordinator of Interventions for Healthy Environment of the World Health Organization, conversation with author, Oxford, UK, January 31, 2017.
6. Yana Jin, Henrik Anderson and Shiqiu Zhang, "Air Pollution Control Policies in China: A Retrospective and Prospects," *International Journal of Environmental Research and Public Health* 13 (12), no. 1219 (December, 2016).
7. Qing Hu, Li Xiaoliang, Lin Aijun, et al., "Total Emission Control Policy in China," *Environmental Development* 25 (March, 2018): 126–129. www.sciencedirect.com/science/article/pii/S2211464517300751.
8. Zhang Chun, "Has China's Impact Assessment Law Lost its Teeth?" *chinadialogue*, July 20, 2016. www.chinadialogue.net/blog/9122-Has-China-s-impact-assessment-law-lost-its-teeth-/en.
9. Chang Jiwen, "Jiance tizhi gaige heyi qiaodong daju" ("How Monitoring System Reform Can Have a Big Impact"), *Zhongguo shengtai wenming* (*China Ecological Civilization*), February 2, 2018. http://huanbao.bjx.com.cn/news/20180202/878616.shtm.
10. "Guanyu zai woguo quanmian tuixing shui wuran wu paifang xukezheng zhidu de jianyi" ("Suggestions on comprehensively implementing water pollutant discharge permit system in our country"), Brief No. 112, Office of the Special Program on Water Pollution Control and Governance Technologies, March 31, 2014. http://nwpcp.mep.gov.cn/zxjz/201403/t20140331_269884.html.
11. Elizabeth Economy, *The River Runs Black: The Environmental Challenge to China's Future* (Ithaca, NY: Cornell University Press).
12. "Cong wumai dao lantian Beijing de kongqi daodi shi zenme bianhao de" ("From Smog to Blue Sky: How Beijing Improved its Air Quality"), December 26, 2017. www.shushiw.com/hangyexinwen/kongjingxilie/xinfengxitong/2017–12-26/4035.html.

13. Abigail R. Jahiel, "The Organization of Environmental Protection in China," *China Quarterly*, no. 156 (December, 1998), 758.
14. See Kenneth Lieberthal, "China's Governing System and Its Impact on Environmental Policy Implementation," *China Environment Series* 1, The Woodrow Wilson Center, 1997.
15. Kenneth G. Lieberthal and David M. Lampton (eds.), *Bureaucracy, Politics, and Decision Making in Post-Mao China* (Berkeley: University of California Press, 1992).
16. "China Is Set to Lose 2% of GDP Cleaning Up Pollution," *Bloomberg Business*, September 17, 2010. www.bloomberg.com/news/articles/2010–09-16/china-set-to-lose-2-of-gdp-fighting-pollution-as-doing-nothing-costs-more.
17. Xi Penghui and Liang Ruobing, "Kongqi wuran dui difang huanbao touru de yingxiang" ("Can Air Pollution Influence the Local Environmental Protection Expenditure"), *Tongji yanjiu* (*Statistical Research*) 32, no. 9 (September, 2015): 76–83.
18. Chinese Academy for Environmental Planning, *Zhongyao huanjing juece cankao* (Chinese Reference for Environmental Decision-making) 12, no. 2, January 29, 2016, 2.
19. Xu Chenggang, "The Fundamental Institutions of China's Reforms and Development," *Journal of Economic Literature* 49, no. 4) (2011): 1076–1151; Victor Shih, Christopher Adolph and Mingxing Liu,"Getting Ahead in the Communist Party: Explaining the Advancement of Central Committee Members in China," *American Political Science Review* 106, no. 1 (February, 2012): 166–187; Xiaobo Lü and Pierre F. Landry, "Show Me the Money: Interjurisdiction Political Competition and Fiscal Extraction in China," *American Political Science Review* 108, no. 3 (August, 2014): 706–722.
20. See Jing Wu, Yongheng Deng, Jun Huang, et al., "Incentives and Outcomes: China's Environmental Policy," ECGI – Finance Working Paper No. 368, (July, 2013). https://ssrn.com/abstract=2206043 or http://dx.doi.org/10.2139/ssrn.2206043.
21. Zhang Liwei, "Gansu huixian jiti qianzhongdu shijian diaocha" ("Investigation of Collective Lead Poisoning in Hui County, Gansu Province"), *21 shiji jingji baodao* (*Financial Report of the 21st Century*), September 12, 2006. http://news.sina.com.cn/c/2006-09-12/104710988048.shtml.
22. "Investigation of Cadmium Pollution in Liuyang, Hunan," *Nanfang dushi bao* (*Southern Metropolis Daily*), August 11, 2009. http://news.sina.com.cn/c/sd/2009-08-11/055918407427.shtml
23. Ma Tianjie, "China's Move to Centralize Environmental Oversight," *The Diplomat*, November 25, 2015. https://thediplomat.com/2015/11/chinas-move-to-centralize-environmental-oversight/.
24. Cao Hongyan,"Zhuanxiang zhiwu datouru chengxiao jihe" ("What Is the Result of Big Investment in Special Pollution Control Programs"), *Jingji ribao* (*Economic Daily*), February 21, 2017. www.mof.gov.cn/zhengwuxinxi/caijingshidian/jjrb/201702/t20170221_2538658.html.
25. Liu Youbin, MEP spokesperson, Remarks. See He Yingchun and Dong Jing (eds.), "Huanbaobu: Jiang mingque daying lantian baoweizhan de shijianbiao he luxiantu" ("MEP Will Explicate Timetable and Road Map on Winning the Battle of Protecting Blue Sky"), *Renmin wang*, December 28, 2017. http://env.people.com.cn/n1/2017/1228/c1010-29734538.html.

26. National Development and Reform Commission, "Nengyuan fazhan shisanwu guihua" ("The Thirteenth Five Year Plan on Energy Development"), December 2016. www.ndrc.gov.cn/zcfb/zcfbghwb/201701/W020170117350627940556.pdf.
27. Wang Peng, "Cong qinghua xiaozhang dao huanbao buzhang" ("From Tsinghua University President to Minister of Environmental Protection"), *Souhu*, March 7, 2015. http://news.sohu.com/20150307/n409450501.shtml.
28. The State Council, "The Action Plan for Prevention and Control of Water Pollution," Regulatory Documents of the State Council, April 2, 2015. www.lawinfochina.com/display.aspx?id=18897&lib=law.
29. Christopher Beam, "China Tries a New Tactic to Combat Pollution: Transparency," *The New Yorker*, February 6, 2015.
30. "Clean Air Action Plan: The Way Forward," *Greenpeace*, February 2016. www.greenpeace.org/eastasia/Global/eastasia/publications/reports/climate-energy/2016/Clean%20Air%20Action%20Plan,%20The%20way%20forward.pdf.
31. Chen Yongqing, deputy inspector for the MEP's Department of Pollution Prevention and Control, Remarks quoted in Zhang Chun, "China Trials Environmental Audits to Hold Officials to Account," *chinadialogue*, June 18, 2015. www.chinadialogue.net/article/show/single/en/7990-China-trials-environmental-audits-to-hold-officials-to-account.
32. The Office of Beijing Municipal Government, "Beijing shi 2013–2017 nian qingjie kongqi xingdong jihua zhongdian renwu fenjie 2017 nian gongzuo cuoshi" ("Measures of the Breakdown of Priority Tasks for Year 2017 in Implementing the 2013–2017 Cleaning Air Action Plan in Beijing"), January 20, 2017. http://zhengce.beijing.gov.cn/library/192/33/401/2751/1242131/142721/74831.pdf.
33. "Chairman of Everything," *The Economist*, April 2, 2016. www.economist.com/china/2016/04/02/chairman-of-everything.
34. Professor Zheng Yongnian, interview with author, Singapore, November 16, 2017.
35. Chinese scholar, conversation with author, Beijing, November 17, 2018.
36. Julia Bowie and David Gitter, "Abroad or at Home, China Puts Party First," *Foreign Policy*, December 5, 2018. https://foreignpolicy.com/2018/12/05/abroad-or-at-home-china-puts-party-first-global-influence-united-front/.
37. "Renda bimu: Xi Jinping yanjiang qiangdiao dang lingdao yiqie" ("NPC Came to a Close: Xi Jinping's Speech Emphasized, 'The Party leads everything'"), *BBC*, March 10, 2018. www.bbc.com/zhongwen/simp/chinese-news-43468026.
38. Duan Min, "Siguanhui jinzhu xizhang simiao shou sengren zhenxin huanying" ("Monks Sincerely Welcome the Presence of Temple Management Committee"), *Xizang ribao* (*Tibet Times*), February 14, 2012. http://news.ifeng.com/mainland/detail_2012_02/14/12485449_0.shtml; "China to Strengthen Communist Party's Role in Non-govt Bodies," *Reuters*, August 21, 2016, www.reuters.com/article/us-china-ngos/china-to-strengthen-communist-partys-role-in-non-govt-bodies-idUSKCN10X07C; Zhang Lin, "Chinese Communist Party Needs to Curtail its Presence in Private Businesses," *South China Morning Post*, November 26, 2018. www.scmp.com/economy/china-economy/article/2174811/chinese-communist-party-needs-curtail-its-presence-private.

39. Wang Xiangwei, "As Fears of Xi's Personality Cult Deepen, China Must Dial Down the Propaganda," *South China Morning Post*, March 24, 2018. www.scmp.com/week-asia/opinion/article/2138481/fears-xis-personality-cult-deepen-china-must-dial-down-propaganda.
40. Xinhua News, "Ruhe lijie quandang bixu laogu shuli zhengzhi yishi daju yishi hexin yish kanqi yishi" ("How to understand that the whole party must firmly establish political consciousness, overall situation consciousness, core consciousness"), Xinhua, November 28, 2016. www.xinhuanet.com/politics/2016-11/28/c_1120002035.htm.
41. Li Huan, "Tianjin shiwei gongzuo huiyi zhaokai" ("Tianjin Municipal Party Committee Meeting Opened"), *Tianjin ribao* (*Tianjin Daily*), October 21, 2016. www.chinanews.com/gn/2016/10-21/8039279.shtml.
42. Austin Ramzy, "Ousted Chinese Official is Accused of Plotting against Communist Party," *New York Times*, October 20, 2017. www.nytimes.com/2017/10/20/world/asia/china-sun-zhengcai-disgraced.html.
43. BBC, "Charting China's 'Great Purge' under Xi," *BBC News*, October 23, 2017. www.bbc.com/news/world-asia-china-41670162.
44. Yuan Ni, "The list of high-ranking officials at the ministerial level and above in the 18th and 19th National Congress of the Republic of China is the largest (Table)," Economic Daily – China Economic Net, October 24, 2019. http://district.ce.cn/newarea/sddy/201410/03/t20141003_3638299.shtml.
45. "32 Provincial Level Party Heads Pledge to Comprehensively Strengthen Party Discipline Prior to the 19th Party Congress," *Renmin ribao* (*People's Daily*), September 20, 2017. http://mini.eastday.com/a/170920085303080-4.html.
46. According to a conservative estimate made by Professor He Jiahong of Renmin University, China has at least two million corrupt officials not being investigated, but just investigating and convicting them would take forty to fifty years. Sun Riwen (ed.), "He Jiahong: Teshe tanguan shi tupo fanfu guaiquan de chulu" ("He Jiahong: Pardoning Corrupted Official Provides a Solution to the Anti-corruption Dilemma"), Renmin University School of Law, May 23, 2015. www.law.ruc.edu.cn/article/?48624.html.
47. Yanzhong Huang, "The Anticorruption Campaign and Rising Suicides in China's Officialdom," *Asia Unbound*, November 25, 2014. www.cfr.org/blog/anticorruption-campaign-and-rising-suicides-chinas-officialdom.
48. Sina News, "Bufeng guanyuan yin fanfu dui qiye ruan jujue" ("Some Government Officials Said Soft "No"'s to Business Firms due to the Anti-corruption Campaign"), *siyuefeng.com*, March 27, 2015. http://fenxiang.siyuefeng.com/article/10643.
49. Fu Xu, "Li Keqiang: Guowuyuan jue bu fa kongtou wenjian" ("Li Keqiang: The State Council Should Never Issue an Empty Document"), China Government Network, May 30, 2014. www.gov.cn/xinwen/2014-05/30/content_2691109.htm.
50. Wang Xiaolu, "Investigation Suggests that Grey Income in Our Country Exceeds Six Trillion, Accounting for 12% of GDP," *Xin shi ji* (*New Century*), September 23, 2013. https://finance.qq.com/a/20130923/000044.htm.

51. "Civil service jobs proving less attractive," *Shanghai Daily*, April 9, 2015. http://china.org.cn/china/2015-04/09/content_35274407.htm.
52. Friend in Shanghai, phone interview with author, April 28, 2018.
53. "Gaige Zhong de xin wenti" ("New Problems in Reform"), *Zhonggai yanjiu* (*China Reform Research*), July 1, 2018. https://mp.weixin.qq.com/s/N9sAGhogj1wIXJyA4_QH9w.
54. Private Wechat discussion, October 16, 2016.
55. "Huanbao ducha Fujian xingdong" (Operation of Environmental Protection Supervision in Fujian), *Diyi caijing* (*yicai*), October 30, 2016. www.yicai.com/news/5146166.html.
56. Local head of environmental protection bureau, interview with author, Zhenjiang, December 18, 2015.
57. Lily Kuo, "China 'Environment Census' Reveals 50% Rise in Pollution Sources," *The Guardian*, March 30, 2018. www.theguardian.com/world/2018/mar/31/china-environment-census-reveals-50-rise-in-pollution-sources.
58. Chang Jiwen, "Ruhe pojie huanjing baohu de xingshi zhuyi" ("How to Address Formalism in Environmental Protection"), *Sina*, May 24, 2017. http://finance.sina.com.cn/wm/2017-05-24/doc-ifyfkqwe0903451.shtml.
59. Chang Jiwen, "Huanjing baohu xu dangzheng tongze" ("Environmental Protection Requires Party-administration Shared Responsibility"), *Huanwei zhisheng* (*Voice of Environmental Sanitation*), September 6, 2017. http://huanbao.bjx.com.cn/news/20170906/848144-2.shtml.
60. Cao Xiaojia, "Who Can Become a Powerful Head of Environmental Protection Bureau?" *Zhongguo huanjing bao* (*China Environmental Daily*), December 17, 2014. http://news.ifeng.com/a/20141218/42747354_0.shtml.
61. Liu Jia, "Beijing yinglai huanbao dai shizhang" ("Beijing Welcomes 'Acting Mayor of Environmental Protection"), *Nanfang zhoumo* (*Southern Weekend*), June 3, 2017. www.infzm.com/content/125016.
62. Li Zhiqing, "Building Ecological Civilization Remains the Main Problem in the Development Situation of Contemporary Environmental Economy" ("Shengtai wenming jianshe rengshi dangqian huanjing jingji fazhan xingshi de zuizhuiyao maodun"), *Souhu*, May 27, 2017. http://lizhiqingsh.blog.sohu.com/324433409.html.
63. See David Victor and Kal Raustiala, "The Regime Complex for Plant Genetic Resources," *International Organization* 32, no. 2, (April, 2004): 147–154.
64. Ruth W. Grant and Robert O. Keohane, "Accountability and Abuse of Power in World Politics," *American Political Science Review* 99, no. 1 (February, 2005): 37
65. Zhang, "China Trials Environmental Audits to Hold Officials to Account."
66. "Coal Burning No Longer Major Source of Beijing PM2.5: Study," Xinhua, May 14, 2018. www.chinadaily.com.cn/a/201805/14/WS5af94c18a3103f6866ee8415.html.
67. Yanzhong Huang, "Tackling China's Environmental Health Crisis," Expert Brief, Council on Foreign Relations, May 14, 2015. www.cfr.org/expert-brief/tackling-chinas-environmental-health-crisis.

68. Ministry of Ecology and Environment, "Policy Areas that Require the MEP to Work with Other Central Agencies," June 1, 2016. www.mee.gov.cn/home/ztbd/rdzl/trfz/ss/201606/t20160601_353132.shtml.
69. Bo Zhiyue, "China's New Environmental Minister: A Rising Star," *The Diplomat*, March 4, 2015. https://thediplomat.com/2015/03/chinas-new-environmental-minister-a-rising-star/.
70. "Huanbaobu buzhang Chen Jining: Guoqu huanbao bu shoufa shi changtai" ("Minister of Environment Chen Jining: Noncompliance in Environmental Protection Laws Was the Normal State"), *Jinhua shibao* (*Beijing Times*), March 2, 2015. http://news.xinhuanet.com/legal/2015-03/02/c_127531601.htm.
71. Sun Lirong, "Linyi zhiwu jizhuanwan" ("Pollution Control Took a Sharp Turn in Linyi"), *The Paper*, July 2, 2015. www.thepaper.cn/baidu.jsp?contid=1347676.
72. Professor CY, interview with author, Shanghai, March 8, 2017.
73. "Jiang gaige jinxing daodi" ("To Follow through on the Reform"), cctv.com. http://tv.cntv.cn/video/C16624/dc2c371a868949c991c38a04385094fe.
74. "Be Reliable in Transmitting the Pressure of Environmental Protection Supervision" *Renmin ribao (People's Daily)*, August 5, 2017. www.xinhuanet.com/comments/2017-08/05/c_1121435430.htm.
75. Wang Tao and Yang Mengqing, "Difang zhengfu gongzuo baogao li de huanbao mimi" ("Environmental Protection Secrets in Local Government Work Reports"), *Nanfang zhoumo* (*Southern Weekend*), March 2, 2017. www.infzm.com/content/123264.
76. Denise Van der Kamp, Peter Lorentzen and Daniel Mattingly, "Racing to the Bottom or to the Top? Decentralization, Revenue Pressures, and Governance Reform in China," *World Development* (July, 2017). https://doi.org/10.1016/j.worlddev.2017.02.021
77. Chenxiang Jing, "Huanjing zhili, zhengfu yaoneng huaqian hui huaqian" ("Environmental Governance: Government Should be Capable of Spending Money and Know How to Spend the Money"), *Qilu wanbao* (*Qilu Evening News*), July 8, 2014. https://weibo.com/p/1001603730092420367206.
78. Chen Jie, "Tenggeli shamo diqu wuran wenti zhaipai" ("Delisting the Pollution Problem in Tengger Desert"), *Xin jing bao* (*The Beijing News*), November 15, 2015. www.guancha.cn/politics/2015_11_15_341350.shtml; Lv Zhongmei, "Kaiqi zhongguo huanjing ziyuan sifa xin zhengcheng" ("Launching the New Journey of China's Environmental Resource Administration of Justice"), November 28, 2017. https://xw.qq.com/cmsid/20171128A0DKTC/20171128A0DKTC00.
79. Huang Han, "Zhibiao zhili jiqi kunjing" ("Performance Target Management and Its Dilemma"), *Haerbing gongye daxue xuebao* (*Journal of Harbin Institute of Technology*), Social Sciences Edition, 18, no. 6 (December 2016), 37–45.
80. The Office of Beijing Municipal Government, "Beijing shi 2013–2017 nian qingjie kongqi xingdong jihua zhongdian renwu fenjie 2017 nian gongzuo cuoshi" ("Measures of the Breakdown of Priority Tasks for Year 2017 in Implementing the 2013–2017 Cleaning Air Action Plan in Beijing").

81. Liu Jia, Yuan Jialu, and Luo Yijue, "Daxian yiguo, chengji kanyou" ("The Deadline Has Passed, the Performance is Worrying"), *Nanfang zhoumo* (*Southern Weekend*), February 1, 2018. www.infzm.com/content/133046.
82. See Xing Li, Chong Liu, Xi Weng, and Li-An Zhou, "Target Setting in Tournaments: Theory and Evidence from China," September 19, 2018. https://ssrn.com/abstract=2937195 or http://dx.doi.org/10.2139/ssrn.2937195.
83. Ma Xinping, "Luoshi zhongyang huanbao juece burong da zhekou" ("No Discount Should Be Allowed in Implementing Central Environmental Protection Policy"), *Zhongguo huanjing bao* (*China Environmental News*), April 13, 2017. www.cenews.com.cn/syyw/201704/t20170413_828086.html.
84. Zhang Yan, "Huanbao ducha de yali chuandao" ("Pressure Transmission in Environmental Protection Supervision"), *The Paper*, April 14, 2017. www.thepaper.cn/newsDetail_forward_1662757.
85. Zhou Li-an, "Zhongguo defang guanyuan de jinsheng jinbiaosai moshi yanjiu" ("Governing China's Local Officials: An Analysis of Promotion Tournament Model"), *Jingji yanjiu* (*Economci Research Journal*), no. 7 (2017): 36–50.
86. See Shiuh-Shen Chien, and Dong-Li Hong, "River leaders in China: Party-state Hierarchy and Transboundary Governance," *Political Geography* 62 (2018), 58–67.
87. Mao Jinglai, "Huanjing jiance shuju shui zaojia shui danze" ("Whoever Found to be Faking Environmental Monitoring Data Should Be Held Responsible"), *Guangming ribao* (*Enlightenment Daily*), September 2, 2017, 9. http://epaper.gmw.cn/gmrb/html/2017-09/02/nw.D110000gmrb_20170902_2-09.htm?div=-1; "Jincheng chengshi kongqi zhiliang jiancequan yi shangshou" ("90% of the Cities' Air Quality Monitoring Authority Has Been Centralized"), *Fazhi ribao* (*Legal Daily*), November 9, 2016. www.ce.cn/xwzx/gnsz/gdxw/201611/09/t20161109_17631867.shtml.
88. Zhao Haoming, Wu Changfeng, and Zhang Yong, "Zhendui bushiying jiakuai quyu ducha zhuanxing fazhan" ("Against the Backdrop of Inadaptation, Accelerate the Development of Regional Supervision Transition"), *Zhongguo huanbao wang*, July 29, 2015. www.chinaenvironment.com/view/ViewNews.aspx?k=20150729100055435
89. Local Environmental Protection Bureau, interview with author, Zhenjiang, China, December 18, 2015.
90. Ma, "China's Move to Centralize Environmental Oversight."
91. "Zhumadian pingyuxian yin huanbao guanli buli duoming guanyuan bei wenze jiangji" ("Several Officials in Pingyu, Zhumadian Were Held Accountable and Demoted for Poor Environmental Protection Mismanagement"), *Fenghuang wang*, March 30, 2015. http://news.ifeng.com/a/20150330/43444306_0.shtml.
92. Wei Hui and Gong Zhihong, "Jingjinji shengtai yitihua qiangshi tuijin" ("A Strong Push for Ecological Integration in the Beijing-Tianjin-Hebei Region"), *Xinhua*, October 10, 2017. www.xinhuanet.com/2017-10/10/c_1121778921.htm.
93. Liu, "Beijing yinglai huanbao dai shizhang" (Beijing Welcomes 'Acting Mayor of Environmental Protection).

94. Michael Forsythe, "China Cancels 103 Coal Plants, Mindful of Smog and Wasted Capacity," *New York Times*, January 18, 2017. www.nytimes.com/2017/01/18/world/asia/china-coal-power-plants-pollution.html.
95. For the legal status of environmental police, see Michael Wunderlich, "Structure and Law Enforcement of Environmental Police in China," *Cambridge Journal of China Studies* 12, no. 4 (2017): 33–49.
96. David Stanway, "Beijing Vows Deep Cut in Coal Use in 2017 to Fight Smog," *Scientific American*, February 6, 2017. www.scientificamerican.com/article/beijing-vows-deep-cut-in-coal-use-in-2017-to-fight-smog/.
97. Cao, "Zhuanxiang zhiwu datouru chengxiao jihe" ("What is the Result of Big Investment in Special Pollution Control Programs.")
98. Ministry of Environmental Protection, et al., *Jingjinji ji zhoubian diaqu 2017–2018 nian qiudongji daqi wuran zonghe zhili gongjian xingdong fang'an* (The problem-solving action plan on comprehensive treatment of air pollution during fall and winter for 2017–2018 in Beijing-Tianjin-Hebei and surrounding regions), August 21, 2017. www.mee.gov.cn/gkml/hbb/bwj/201708/W020170824378273815892.pdf; Zheng Yiwen, "Zhiji sanmei zhili tongdian" ("Facing the Pain Points in Bulk Coal Management"), Fenghuang International Think Tank, November 27, 2017. https://pit.ifeng.com/a/20171127/53627674_0.shtml.
99. Qie Jianrong, "Huanbao dangzheng tongze, yigang shuangze zheng shixian zhidu pobing" ("Institutional Icebreaking for Party-administration Shared Responsibility and One-post-two-responsibilities in Environmental Protection is Underway"), *Fazhi ribao* (*Legal Daily*), January 10, 2018. www.chinanews.com/gn/2018/01-10/8420206.shtml.
100. "130 ge tingguan tongshi bei wenze daodi weisha" ("Why Were 130 Department-Level Officials Held Accountable Simultaneously?"), *Renmin ribao* (*People's Daily*), November 20, 2017. http://politics.people.com.cn/n1/2017/1120/c1001-29656462.html.
101. Chang Jiwen, "Huanbao zhuize wenze" ("Accountability in Environmental Protection"), Souhu, November 17, 2017. www.sohu.com/a/204966715_378134.
102. Xia Kedao, "Bei zhongyang tongbao piping gansu fanle shenmeshi" ("Why Was Gansu Criticized by the Party Center in a Circulated Notice?"), *Haiwai wang (Overseas Network)*, July 21, 2017. http://opinion.haiwainet.cn/n/2017/0721/c456317-31032226.html.
103. David Stanway, "China Launches New Year-long Inspection into Air Pollution in North," *Reuters*, April 5, 2017. www.reuters.com/article/us-china-pollution/china-launches-new-year-long-inspection-into-air-pollution-in-north-idUSKBN178098.
104. Zhang Ke, "Jingjinji sanluanwu qiye zhengzhi jianxiao" ("Regulation of Beijing-Tianjin-Hebei 'Chaotic, Scattered and Polluted' Enterprises is Taking Effect"), *Diyi caijing (yicai)*, November 21, 2017. www.yicai.com/news/5374964.html.
105. He Yingchun, Chu Zirui, "Dahao sanda zhanyi jianshe tianlian shuiqing dijing de meili Zhong guo" ("Complete Three Tasks, and Build a Beautiful China with Blue Sky, Clean Water and Clean Land"), *Renmin wang*, September 25, 2017. http://env.people.com.cn/n1/2017/0925/c1010-29558119.html.

106. "Zhongyang huanbao ducha shai 31 shengfen wenti qingdan" (Central Environmental Protection Supervision Group Unveils List of Issues for 31 Provinces), January 8, 2018. http://wemedia.ifeng.com/44178706/wemedia.shtml.
107. Deng Qi, "Xiang wumai xuanzhan, Beijing chao changgui shouduan zhili daqi" ("Declaring War on Smog: Beijing Adopts Extraordinary Measures in Controlling Air Pollution"), *Xinjing bao* (*Beijing News*), February 26, 2019. www.bjnews.com.cn/feature/2019/02/26/550335.html.
108. Maggie Zhang, "New Environment Tax Will Hit Businesses in China Hard, Say Experts," *South China Morning Post*, October 3, 2017. www.scmp.com/business/china-business/article/2113650/new-environment-tax-will-hit-businesses-china-hard-say.
109. Tianjie Ma and Qin Liu, "China Reshapes Ministries to Better Protect Environment," *chinadialogue*, March 14, 2018. www.chinadialogue.net/article/show/single/en/10502-China-reshapes-ministries-to-better-protect-environment.
110. Zhang Ke, "Jiangsu gongjian daqi wuran fangzhi" ("Jiangsu Tackles Air Pollution Prevention and Control"), *Diyi caijing (yicai)*, August 23, 2019. www.yicai.com/news/100305531.html.
111. "1965: Mao Zedong Gave Birth to a Barefoot Doctor," *Beijing ribao* (*Beijing Daily*), September 25, 2009. http://finance.ifeng.com/news/20090925/1284328.shtml.
112. Dorcas Wong, "Shanghai Businesses to Comply with New Waste Management Norms from July 1," *China Briefing*, June 25, 2019. www.china-briefing.com/news/shanghai-waste-management-china-july-1/.
113. "The Era of Compulsory Garbage Sorting Begins," *China Daily*, June 24, 2019. www.chinadaily.com.cn/a/201906/24/WS5d10650ba3103dbf14329e23.html.

AN ASSESSMENT OF POLICY EFFECTIVENESS

1. State Council, "Daqi wuran fangzhi xingdong jihua" ("The Air Pollution Prevention and Control Action Plan"), *Central Government Portal*, September 10, 2013. www.gov.cn/zwgk/2013–09/12/content_2486773.htm.
2. Prefecture is an administrative division in China that ranks below a province and above a county or county-level city.
3. CCTV, "Shouzhu lushui qingshan" ("Protect the Green Water and Green Mountains"), July 22, 2017. http://m.news.cctv.com/2017/07/22/ARTIPKcW2uNDICbqDICM12rr170722.shtml; Interface China, "Huanbaobu: qunian quanguo 338 ge diji yishang chengshi zhiyou 99 ge kongqi zhiliang dabiao" ("MEP: Only 99 of the 338 Cities at or above the Prefectural Level Meet the Air Quality Standards in the Past Year"), *Souhu*, February 27, 2018. www.sohu.com/a/224297104_313745.
4. Zhang Ke, "Zhongguo gaishan kongqi zhiliang de moshi chulaile" ("China's Model of Improving Air Quality Came Out"), *Diyi Caijing (yicai)*, February 2, 2018. www.yicai.com/news/5397738.html.
5. Zhang, "Zhongguo gaishan kongqi zhiliang de moshi chulaile" (China's Model of Improving Air Quality Came Out).

6. "PM2.5 in Beijing down 54%, but nationwide air quality improvements slow as coal use increases," Greenpeace East Asia. Press release, January 11, 2018. www.greenpeace.org/eastasia/press/releases/climate-energy/2018/PM25-in-Beijing-down-54-nationwide-air-quality-improvements-slow-as-coal-use-increases/.
7. Michael Greenstone, "Four Years after Declaring War on Pollution, China Is Winning," *New York Times*, March 12, 2018. www.nytimes.com/2018/03/12/upshot/china-pollution-environment-longer-lives.html.
8. Yixuan Zheng, Tao Xue, Qiang Zhang, et al., "Air Quality Improvements and Health Benefits from China's Clean Air Action since 2013," *Environmental Research Letters* 12, no. 11 (November, 2017). http://iopscience.iop.org/article/10.1088/1748–9326/aa8a32/meta.
9. Steven Lee Myers, "A Blue Sky in Beijing? It's Not a Fluke, Says Greenpeace," *New York Times*, January 11, 2018. www.nytimes.com/2018/01/11/world/asia/pollution-beijing-declines.html
10. Greenstone, "Four Years after Declaring War on Pollution, China Is Winning."
11. Yawen Chen and Ryan Woo, "Another Chinese City Admits 'Fake' Economic Data," *Reuters*, January 17, 2018. www.reuters.com/article/us-china-economy-data/another-chinese-city-admits-fake-economic-data-idUSKBN1F60I1.
12. David Stanway, "China Punishes Officials for Tampering with Smog Monitoring," *Reuters*, January 14, 2018. www.reuters.com/article/us-china-pollution/china-punishes-officials-for-tampering-with-smog-monitoring-idUSKBN1F402Y; "Penshui saimiansha … zhongguo defang guanyuan taobi kongqi jiance you qizhao" (Spraying Water, Stuffing Cotton Yarn … Chinese Local Officials Have Neat Tricks to Avoid Air Monitoring), *BBC News*, January 15, 2018. www.bbc.com/zhongwen/simp/chinese-news-42687637.
13. David Stanway, "China Environment Officials in Xian Detained for Data Fraud: Xinhua," *Reuters*, October 26, 2016. www.reuters.com/article/us-china-environment/china-environment-officials-in-xian-detained-for-data-fraud-xinhua-idUSKCN12Q09T; Chen Junsong, "Huanjing jiance shuju zaojia" ("Faking Environmental Monitoring Data"), *Zhongguo xinwen wang*, August 6, 2018. www.xinhuanet.com/politics/2018–08/06/c_1123229150.htm.
14. John McGarrity, "China Promises Crackdown on Fake Air Quality Data," *The Third Pole*, April 13, 2015. www.thethirdpole.net/en/2015/04/13/china-promises-crackdown-on-fake-air-quality-data/.
15. Huang Han, "Zhibiao zhili jiqi kunjing" ("Performance Target Management and Its Dilemma"), *Haerbing gongye daxue xuebao* (*Journal of Harbin Institute of Technology*), Social Sciences Edition, 18, no. 6 (December 2016), 37–45.
16. "Yong dashuju fenxi baozheng jiance shuju zhenzhunquan" ("Using Big Data Analysis to Ensure the Reliability, Accuracy and Comprehensiveness of Monitoring Data"), *Zhongguo huanjing bao* (*China Environment News*), July 12, 2018, p. 3. http://news.cenews.com.cn/html/2018–07/12/content_74329.htm.
17. Wang Jin, "Duzheng wenze, qiangli zhiwu" ("Supervision, Accountability, and Strong Pollution Control"), *Fenghuang zhoukan (Phoenix Weekly)*, December 25, 2017. www.ifengweekly.com/detil.php?id=4959.

18. Wang Erde, "Beijing shi dai shizhang Cai Qi: Jinnian tui shitiao jucuo tiewan zhimai" ("Beijing Acting Mayor Cai Qi: Ten Measures Will Be Implemented to Control Haze with an Iron Hand this Year"), *21 shiji jingji baodao* (*Financial Report of the 21st Century*), January 7, 2017. https://m.21jingji.com/article/20170107/herald/5b7a94ef1fd50cd874aa46a98b46ccc2.html.
19. "Renmin ribao qi wen wumai" ("Seven Questions on Haze Raised by the People's Daily"), Renmin ribao kehu duan, January 5, 2017. http://finance.sina.com.cn/china/2017–01-05/doc-ifxzkfuk2318264.shtml.
20. "Renmin ribao qi wen wumai" ("Seven Questions on Haze Raised by the People's Daily").
21. Zachary Wendling, Jay Emerson, Daniel Esty, et al., "China Country Profile," *Environmental Performance Index*. New Haven, CT: Yale Center for Environmental Law & Policy (2018). https://epi.envirocenter.yale.edu/sites/default/files/2018-chn.pdf.
22. "Renmin ribao qi wen wumai" (Seven Questions on Haze Raised by the People's Daily).
23. Steven Lee Myers, "A Blue Sky in Beijing? It's Not a Fluke, Says Greenpeace," *New York Times*, January 11, 2018. www.nytimes.com/2018/01/11/world/asia/pollution-beijing-declines.html.
24. Liu Shixin, "Daying lantian baoweizhan de diqi conghe erlai" ("Where the Confidence of Winning the Blue Sky Protection Campaign is From"), *zhongqing zaixian*, February 5, 2018. http://china.chinadaily.com.cn/2018–02/05/content_35645501.htm.
25. Benjamin van Rooij, Qiaoqiao Zhu, Li Na, and Wang Qiliang, "Centralizing Trends and Pollution Law Enforcement in China," *The China Quarterly*, 231 (September, 2017): 583–606.
26. Yixuan Zheng, Tao Xue, Qiang Zhang, et al., "Air Quality Improvements and Health Benefits from China's Clean Air Action since 2013," *Environmental Research Letters* 12, no. 11 (November, 2017). http://iopscience.iop.org/article/10.1088/1748–9326/aa8a32/meta; Asian Scientist Newsroom, "China On Track to Achieving Air Quality Targets," *Asian Scientist*, November 14, 2017. www.asianscientist.com/2017/11/in-the-lab/china-air-pollution-public-health/.
27. Wendling, Emerson, Esty, et al., "China Country Profile."
28. Xinhua, "Beijing-Tianjin-Hebei Region to Suffer from Ozone Pollution," *China Daily*, May 19, 2017. www.chinadaily.com.cn/china/2017–05/19/content_29414237.htm. See also Ke Li, Daniel J. Jacob, Hong Liao, Lu Shen, et al., "Anthropogenic Drivers of 2013–2017 Trends in Summer Surface Ozone in China," *Proceedings of the National Academy of Sciences* 116 no. 2(January, 2019): 422–427.
29. "PM2.5 in Beijing Down 54%, but Nationwide Air Quality Improvements Slow as Coal Use Increases," Greenpeace East Asia, January 11, 2018. www.greenpeace.org/eastasia/press/releases/climate-energy/2018/PM25-in-Beijing-down-54-nationwide-air-quality-improvements-slow-as-coal-use-increases/.
30. Health Effects Institute, *State of Global Air 2019*, Special Report (Boston, MA: Health Effects Institute, 2019), 6.
31. "Zhongyang huanbao duchazu shai 21 shengfen wenti qingdan" ("Central Environmental Protection Supervision Team Unveils Issue List for 31 Provinces"), January 8, 2018. http://wemedia.ifeng.com/44178706/wemedia.shtml.

32. Wang Tao and Yang Mengqing, "Difang zhengfu gongzuo baogao lide huanbao mimi" ("Environmental Protection Secret in Local Government Work Report"), *Nanfang zhoumo* (*Southern Weekend*), March 2, 2017. www.infzm.com/content/123264.
33. Professor Yang Gonghuan, interview with author, New York, November 20, 2017.
34. GBD MAPS Working Group, *Burden of Disease Attributable to Coal-Burning and Other Major Sources of Air Pollution in China*, Special Report 20 (Boston, MA: Health Effects Institute, 2016).
35. Qing Wang, Jiaonan Wang, Jinhui Zhou, et al., "Estimation of PM2·5-Associated Disease Burden in China in 2020 and 2030 Using Population and Air Quality Scenarios: A Modelling Study," *Lancet Planet Health* 3 (2019): 71–80.
36. Delin Fang, Bin Chen, Klaus Hubacek, et al., "Clean Air for Some: Unintended Spillover Effects of Regional Air Pollution Policies," *Science Advances* 24 (April, 2019): 1–10.
37. Huang, "Zhibiao zhili jiqi kunjing" ("Performance Target Management and Its Dilemma").
38. Michael Mann, *The Sources of Social Power, Vol. 2: The Rise of Classes and Nation States, 1760–1914* (New York, NY: Cambridge University Press, 1993), 59.
39. "Zhengfu gongzuo baogao li de huanbao qiangyin" (A Strong Tone of Environmental Protection in Government Work Reports"), *Xinhua*, March 5, 2017. www.xinhuanet .com//2017–03/05/c_1120572008.htm.
40. Shang Guangxu, et al., "Shisan wu chuchen tuoliu tuoxiuao hangye zhengce daoxiang ji fazhan qushi" ("Policy Directions and Development Trend of Desulphurization and Denitrification Industry under the Thirteenth Five-Year Plan"), China Environmental Protection Industry Association, December 2, 2016. www.cnenergy.org/hb/201612/ t20161202_409388.html.
41. Li He, "Wanyi ji huanbao chanye zaoyu fazhan zhengtong" ("Trillion-yuan Environmental Protection Industry is Experiencing Growing Pain"), *Keji ribao* (*Science and Technology Daily*), July 16, 2018. www.stdaily.com/kjrb/kjrbbm/2018–07/16/ content_690141.shtml.
42. "Zouxiang huanbao qiangyin dahao lantian baoweizhan" ("Play a Strong Tone of Environmental Protection, and Do a Good Job in the Blue Sky Protection Campaign").
43. Den Qi, "Huanbaobu: daqi shitiao kongqi gaishan mubiao neng shixian" ("The Air Quality Improvement Goals Contained in *qi shi tiao* Can Be Achieved"), *Xin jing bao* (*The Beijing News*), December 29, 2017. www.xinhuanet.com/energy/2017–12/29/ c_1122183654.htm.
44. Yang Ye, "Huanbao xiaozu tuji jiancha" ("Environmental Groups' Surprise Inspection"), *Jingji cankao bao* (*Economic Information Daily*), December 2, 2015. http:// news.sciencenet.cn/htmlnews/2015/12/333065.shtm.
45. Gan Lian, "Zhili Yichang wumai daodi yao hua duo shao qian?" ("How Much Does it Cost to Control the Smog"), *Sina News*, November 8, 2016. http://news.sina.com.cn/ pl/2016–11-08/doc-ifxxneua4403105.shtml.
46. Feng Jun, "Qianyi zhimai zhijin huo wugong erfan" ("100 Billion Fund for Smog Control May End Up Achieving Nothing"), *Tenxun caijing lengjing*, December 28, 2015. www.tanpaifang.com/ditanhuanbao/2015/1228/49802.html.

47. When applied to coal burning power plants, ultra-low emission requirements seek to bring emission from these plants down to the level of gas-burning counterparts.
48. Feng Jun, "Qianyi zhimai zhijin huo wugong erfan" ("100 Billion Fund for Smog Control May End Up Achieving Nothing").
49. Feng Jun, "Qianyi zhimai zhijin huo wugong erfan" ("100 Billion Fund for Smog Control May End Up Achieving Nothing").
50. "Huanbaobu: daqi shitiao kongqi gaishan mubiao neng shixian" ("The Air Quality Improvement Goals Contained in *qi shi tiao* Can Be Achieved"), *Xin jing bao* (*The Beijing News*), December 29, 2017. www.xinhuanet.com/energy/2017–12/29/c_1122183654.htm
51. Xinhua, "Coal Burning No Longer Major Source of Beijing PM2.5: Study," *China Daily*, May 14, 2018. www.chinadaily.com.cn/a/201805/14/WS5af94c18a3103f6866ee8415.html.
52. Zhang Ke, "Qiche baoyouliang 27 nian zeng 33 bei" ("Vehicle Ownership Increased 33 Fold in 27 Years"), *Diyi caijing (yicai)*, November 21, 2018. www.yicai.com/news/100064032.html.
53. Greenpeace, "Guonei 12jia chaoji paifang ranmei dianchang shishi paifang qingkuang diaocha" ("An Investigation of Actual Emission Status for 12 Domestic 'Ultra-low Emission' Coal Power Plants"), December 16, 2015.
54. Dong Ruiqiang, "Huanbao ducha zuixin tongbao" ("The Latest Report on Environmental Protection Supervision"), *Jingji guanchao bao (The Economic Observer)*, August 23, 2017. www.eeo.com.cn/2017/0823/311198.shtml.
55. Zhonggong shuilibu dangzu (CCP Ministry of Water Resources Party Group), "Jianshe renshui hexie meili zhongguo" ("Build a Beautiful China with Harmonious Relationship between Human and Water"), *Quishi* (Seeking Truth), September 11, 2017. http://econ.cssn.cn/jjx/xk/jjx_lljjx/rkjjldjjx/201709/t20170911_3636316.shtml.
56. China Water Risk, "2016 State of Environment Report Review," June 14, 2017. www.chinawaterrisk.org/resources/analysis-reviews/2016-state-of-environment-report-review/.
57. Zhang Ke, "Huanbaobu huiying diyi caijing tiwen" ("MEP Answers Questions from diyi caijing"), *Diyi caijing (yicai)*, March 20, 2017. https://m.yicai.com/news/5249962.html.
58. Dong Ruiqiang, "Huanbaobu: quanguo heichou shuiti zhengzhi jin wancheng 44.1%" ("Only 44.1 Percent of the Work of Treating the Black-odor Rivers Has Been Completed"), *Jingji guanchao bao (The Economic Observer)*, August 25, 2017. www.eeo.com.cn/2017/0825/311282.shtml.
59. Shan Jie, "Officials Toss Chemicals in River to Mask Pollution, Waste $6.8 Million," *Global Times*, November 21, 2018. www.globaltimes.cn/content/1128343.shtml.
60. Ban Juanjuan, "Nongcun wushui youxiao zhili anxia kuaijinjian" ("Fast-forward Button Has Been Pressed in Effective Treatment of Polluted Water in the Countryside"), *Jingji cankao bao* (*Economic Information Daily*), June 21, 2018. http://jjckb.xinhuanet.com/2018–06/21/c_137269250.htm.
61. Zuo Shengdan (ed.), "Li Zheshu: On the whole, China's Water Ecological Environment Is Not Optimistic," *China News*, August 23, 2019. www.chinanews.com/gn/2019/08–23/8936293.shtml.

62. Local government officials, interview with author, Zhenjiang, December 13, 2014.
63. "Dong Lihua: 2017 nian woguo jingji pingwen zengzhang" ("Dong Lihua: Our Country's Economy Registered Steady Growth in 2017"), January 19, 2018. www.stats.gov.cn/tjsj/sjjd/201801/t20180119_1575457.html.
64. "Zhongguo jihua dao 2020 nian jiancheng daxing meitan qiye" ("China Plans to Complete the Building of Large-scale Coal Enterprises by 2020"). http://shengyuexintube.com/index/news_show/id/145.html.
65. China Energy Storage Network News, "2016 nian qingjie nengyuan xiaofei bizhong tisheng zhi 19.7%" ("The Share of Clean Energy Consumption Ration Increased to 19.7% in 2016"), *Zhongguo meitan ziyuan wang*, March 4, 2017. www.escn.com.cn/news/show-402004.html.
66. Zhang, "Zhongguo gaishan kongqi zhiliang de moshi chulai le" ("China's Model of Improving Air Quality is Now Available").
67. "Can China's Electric Car Market Keep Growing if Beijing Pulls the Plug on Subsidies?," *South China Morning Post*, June 30, 2018. www.scmp.com/business/companies/article/2153041/can-chinas-electric-car-market-keep-growing-if-beijing-pulls-plug.
68. Li Xiyin, "Kejibu buzhang: dui jinnian xinnengyuan che xiaoliang dadao 100 wanliang chi leguan taidu" ("Minister of Science and Technology is Optimistic toward Selling 1 Million New Energy Vehicles this Year"), *The Paper*, January 20, 2018. https://m.thepaper.cn/newsDetail_forward_1959414?from=timeline&isappinstalled=0.
69. Wang Fang, "Lantian baowei zhan, weihe shi Yichang jiannan er chijiu de gongjianzhan?" ("Blue Sky Protection Campaign: Why is it a Difficult and Lasting Tough Fight?"), March 14, 2018. www.sohu.com/a/225526708_714642.
70. Chinese journalist, private communication with author, January 4, 2017.
71. Christine Shearer, Aiqun Yu, and Ted Nace, "Tsunami Warning: Can China's Central Authorities Stop a Massive Surge in New Coal Plant Caused by Provincial Overpermitting?" CoalSwarm, September 2018. https://endcoal.org/2018/09/tsunami-warning/.
72. "Xu Changming: 2020 nian woguo qiche baoyouliang huo da 2.5 yiliang" ("Xu Changming: China's Car Ownership in 2020 May Reach 250 Million"), Sina, October 21, 2015. http://auto.sina.com.cn/news/hy/2015–10-21/detail-ifxivsce7001015.shtml.
73. Zhang, "Zhongguo gaishan kongqi zhiliang de moshi chulaile" ("China's Model of Improving Air Quality Came Out).
74. Cao Hongyan,"Zhuanxiang zhiwu datouru chengxiao jihe" ("What is the Result of Big Investment in Special Pollution Control Programs"), *Jingji ribao* (*Economic Daily*), February 21, 2017. www.mof.gov.cn/zhengwuxinxi/caijingshidian/jjrb/201702/t20170221_2538658.html.
75. Tianjie Ma, "Beijing Issues First Smog Red Alert for 2016," *chinadialogue*, December 15, 2016. www.chinadialogue.net/blog/9505-Beijing-issues-first-smog-red-alert-for-2–16/en
76. David Stanway, "China's Beijing-Tianjin-Hebei to Set Up Anti-pollution Body: Report," *Reuters*, October 31, 2017. www.reuters.com/article/us-china-pollution-jingjinji/chinas-beijing-tianjin-hebei-to-set-up-anti-pollution-body-report-idUSKBN1D139K.

77. Deng Qi, "Huan bao bu: daqi shitiao kongqi gaishan mubiao neng shixian" ("MEP: Air Improving Targets in 'Ten Measures for air' Can Be Hit"), *Xin jing bao* (*The Beijing News*), December 29, 2017. www.xinhuanet.com/energy/2017-12/29/c_1122183654.htm.
78. Liu Xiaoping (ed.), "Huanbao yu weisheng bumen huoke gongtong kaizhan wuran jiankang yingxiang diaocha" ("Departments of Environmental Protection and Health May Jointly Conduct Survey on the Health Impact of Pollution"), *Fazhi ribao* (*Legal Daily*), March 27, 2017. http://legal.people.com.cn/n1/2017/0327/c42510-29171367.html.
79. Liu, "Huanbao yu weisheng bumen huoke gongtong kaizhan wuran jiankang yingxiang diaocha" ("Departments of Environmental Protection and Health May Jointly Conduct Survey on the Health Impact of Pollution").
80. Dr. Shang Qi, China Center for Disease Control, interview with author, Beijing, July 1, 2015.
81. Professor DT, communication with author, January 8, 2018.
82. Xu Dongqun, "Woguo huanjing yu jiankang mianlin de tiaozhan ji yingdui celue."
83. Jennifer Holdaway, interview with author, Beijing, June 30, 2015; Local government officials from the Taiyuan Environmental Protection Bureau, interview with author, Taiyuan, December 14, 2015.
84. Shan Huan Han, "Shanxi sheng guanyu jiaqiang turang turan wuran zhuangkuang diaocha shuju baomi tongzhi" ("Circular on Strengthening Soil Pollution Survey Data Security, Shaanxi Province"), *Wangyi caijing*, February 10, 2013. http://money.163.com/13/0219/13/8O32QCK100252G50.html.
85. Wu Jiajia, "Guojia weijiwei huiying wumai dui jiankang yingxiang huati" ("NHFPC Responds to Topics on the Health Impact of Smog"), *Zhongguo jingji wang*, January 8, 2017. http://news.sohu.com/20170108/n478096624.shtml.
86. Wu, "Guojia weijiwei huiying wumai dui jiankang yingxiang huati" ("NHFPC Responds to Topics on the Health Impact of Smog").
87. Guillaume Pitron, "China's Army of Green Activists," *Gulf News*, August 11, 2017. http://gulfnews.com/culture/environment/china-s-army-of-green-activists-1.2071769.
88. About chinadialogue, see www.chinadialogue.net/pages/about
89. Feng Yongfeng, interview with author, Beijing, July 1, 2015.
90. "Achievements," Greenpeace East Asia. www.greenpeace.org/eastasia/about/achievements/.
91. Ma Jun, interview with author, Beijing, China, July 1, 2015.
92. "Woguo gelei minjian huanbao zuzhi yi fazhan zhi jiqianjia" ("Our Country Already Had Several Thousand Civilian Environmental Protection Organizations"), *Gonyi shibao* (*Public Benefits Times*), December 5, 2017. www.gongyishibao.com/html/gongyizixun/8125.html.
93. Wu Haoliang and Zhu Lihua, interview with author, He Yi Institute, Beijing, July 3, 2015.
94. Wu Haoliang and Zhu Lihua, interview with author, He Yi Institute, Beijing, July 3, 2015.

95. Xu Nan and Zhang Chun, "Why are China's Anti-pollution Lawsuits Stalling?" *chinadialogue*, June 23, 2015. www.chinadialogue.net/article/show/single/en/7986-Why-are-China-s-anti-pollution-lawsuits-stalling.
96. Xinhua, "Coal Burning No Longer Major Source of Beijing PM2.5: Study," *China Daily*, May 14, 2018. www.chinadaily.com.cn/a/201805/14/WS5af94c18a3103f6866ee8415.html.
97. Tang Qi, "Baogao xi zhongguo huanjing sifa xianzhuang" ("Report Analyzes the Status of China's Environmental Justice"), *Zhongguo xinwen wang*, July 14, 2017. www.chinanews.com/gn/2017/07-14/8277397.shtml.
98. "Huanjing gongyi susong weihe yuleng" ("Why is Environmental Public Interest Litigation Not Popular"), China Government Network, January 24, 2017. http://health.people.com.cn/n1/2017/0124/c14739-29046841.html.
99. "Huanjing gongyi susong weihe yuleng" ("Why is Environmental Public Interest Litigation Not Popular").
100. People's Supreme Court Press Conference, Beijing, March 2, 2019. www.chinacourt.org/article/subjectdetail/id/MzAwNMgqMIABAA%3D%3D.shtml?from=timeline&isappinstalled=0.
101. Yanzhong Huang, "At the Mercy of the State: Health Philanthropy in China," *Voluntas: International Journal of Voluntary and Nonprofit Organizations* no. 30 (2019): 634–646.
102. "Huanjing gongyi susong weihe yuleng" ("Why is Environmental Public Interest Litigation Not Popular").
103. Under the Chinese legal system, the losing party must pay the court fees of the winners as well. Zhang Chun and Tang Damin, "The Changzhou Soil Pollution Case is Far from Over," *chinadialogue*, February 24, 2017. www.chinadialogue.net/article/show/single/en/9630-The-Changzhou-soil-pollution-case-is-far-from-over.
104. William J Schulte and Li Haitang, "Yunnan Chemical Factory Becomes Testing Ground for Citizen Lawsuits," *chinadialogue*, August 23, 2017. www.chinadialogue.net/article/show/single/en/9983-Yunnan-chemical-factory-becomes-testing-ground-for-citizen-lawsuits.
105. Didi Kirsten Tatlow, "China Has Made Strides in Addressing Air Pollution, Environmentalist Says," *New York Times*, December 16, 2016. www.nytimes.com/2016/12/16/world/asia/china-air-pollution-ma-jun.html.
106. Lily Kuo, " 'A Sort of Eco-dictatorship': Shanghai Grapples with Strict New Recycling Laws," *The Guardian*, July 11, 2019. www.theguardian.com/world/2019/jul/12/a-sort-of-eco-dictatorship-shanghai-grapples-with-strict-new-recycling-laws.
107. Wu Yixiu, "Shanghai's Compulsory Waste Sorting Begins," *chinadialogue*, July 2, 2019. www.chinadialogue.net/article/show/single/en/11349-Shanghai-s-compulsory-waste.
108. Li Biao, "2015 nian woguo wuran zhili touzi jin 9000 yi" ("Our Country Invested Nearly 900 Billion in Pollution Control"), meiri jingji xinwen, June 15, 2017. www.nbd.com.cn/articles/2017-06-15/1117584.html.
109. Li, "2015 nian woguo wuran zhili touzi jin 9000 yi" ("Our Country Invested Nearly 900 Billion in Pollution Control").

110. By the end of June 2019, China's local authorities had amassed a total debt of at least 21 trillion *yuan*, about 23 percent of China's GDP in 2018. See Yanzhong Huang, "Why Did One-Quarter of the World's Pigs Die in a Year?" *New York Times*, international edition, January 3, 2020, p. 9. A county-level city in Eastern China has an annual fiscal revenue of 3.2 billion *yuan*, yet it is on debt for 85 billion, and needs to pay annual interest of 5.2 billion. Author's interview, summer 2018.
111. Yanzhong Huang, "Is China Serious about Pollution Controls?" Council on Foreign Relations, November 19, 2015. www.cfr.org/expert-brief/china-serious-about-pollution-controls. See also China Public Private Partnerships Center (CPPPC), project database. www.cpppc.org:8086/pppcentral/map/toPPPChooseList.do.
112. Li Liyun, Li Lina, and Liu Guanghua, "Zhongche qiche jituan chuli guode shenghuo wushui ke yongyu luhua" ("Domestic Sewage Treated by CRRC Qiqihar Rolling Stock Co. is Safe for Afforestation"), *Keji ribao* (*Science and Technology Daily*), August 9, 2018. www.stdaily.com/02/difangyaowen/2018-08/09/content_698692.shtml.
113. "2017 nian zhongguo huanbao hangye PPP moshi chengwei hangye zhuliu moshi" ("PPP Model in China's Environmental Protection Industry Becomes Mainstream Model in the Industry in 2017"), April 27, 2017. www.chyxx.com/industry/201704/517832.html.
114. Andrew Baston, "China Considers Tradable Pollution-Rights Permits," *Washington Post*, March 29, 2006. www.washingtonpost.com/wp-dyn/content/article/2006/03/28/AR2006032801565.html.
115. "Paiququan youchang shiyong he jiaoyi jine xianzhu zengjia" ("A Significant Increase in the Revenue from Paid Use and Transfer of Pollution Rights"), *Jingji ribao* (*Economic Daily*), January 24, 2019. www.gov.cn/xinwen/2019-01/24/content_5360745.htm.
116. Qiao Jinliang, "Turang xiufu, wanyi guimo dangao ruhe fen" ("How to Divide the a Trillion Yuan Cake in Soil Remediation"), *Zhongguo jingji wang*, April 8, 2017. www.ce.cn/xwzx/gnsz/gdxw/201704/08/t20170408_21802247.shtml.
117. Li Biao, "Huanbao PPP zhongbiao dahu nantao changzhai yali" ("Major Tenders of Environmental Protection PPP Face the Pressure of Repaying the Debts"), *Meiri jingji xinwen (Daily Economic News)*, July 22, 2018. www.nbd.com.cn/articles/2018-07-22/1237508.html.
118. Cui Ying and Qian Qingjing, "Woguo paiwuquan jiaoyi shichang de fazhan qingkuang wenti he zhengce jianyi" ("Development, Problems and Policy Recommendations Regarding Our Country's PRT Market"), *Guoji huanbao zaixian (International Environmental Protection Online)*, November 11, 2018. https://m.huanbao-world.com/view.php?aid=55663.
119. Zhu Yan, "Paiwuquan Jiayi shinian tuier bu guang, wenti chuzai nale" ("PRT Has Been Promoted for Ten Years But is Still Not Popular, Why?"), *Zhonggu nengyuan bao* (*China Energy News*), May 22, 2017. www.cnenergy.org/yw/201705/t20170522_444378.html; Cui and Qian, "Woguo paiwuquan jiaoyi shichang de fazhan qingkuang wenti he zhengce jianyi" ("Development, Problems and Policy Recommendations Regarding Our Country's PRT Market").

120. "Paiwuquan jiaoyi youjia wushi xianzhuang jidai gaibian" ("The Situation that Pollution Rights Trade Has a Price But No Sales Should Be Changed), *Jingji cankao bao* (*Economic Information Daily*), June 13, 2016. http://finance.people.com.cn/n1/2016/0613/c1004-28428971.html.
121. Cui and Qian, "Woguo paiwuquan jiaoyi shichang de fazhan qingkuang wenti he zhengce jianyi."
122. "Changjiang xian huagong suojiang kunju" ("Yangtze River is in a Dilemma of Having Chemical Industry Blocking River"), *Jingji cankao bao* (*Economic Information Daily*), June 30, 2016. http://env.people.com.cn/n1/2016/0630/c1010-28510262.html.
123. Bloomberg News, "China Outsourcing Smog to West Region Stirs Protest," *Bloomberg*, March 6, 2014. www.bloomberg.com/opinion/articles/2018-11-15/trump-tweets-against-mueller-he-s-running-scared; Mark Eels, "China's Delicate Pursuit of Natural Gas," *The Diplomat*, April 15, 2014. https://thediplomat.com/2014/04/chinas-delicate-pursuit-of-natural-gas/.
124. Valerie J. Karplus, "Double Impact: Why China Needs Coordinated Air Quality and Climate Strategies," *Paulson Papers On Energy and Environment*, February 2015.
125. Yanzhong Huang and Dali Yang, "Population Control and State Coercion in China," in *Holding China Together: Diversity and National Integration in the Post-Deng Era*, edited by Barry Naughton and Dali Yang, 193–225 (New York: Cambridge University Press, 2004).
126. Wang Wei, "Youfa biyi caineng zhiwu bisheng" (Pollution Control Can Be Successful Only by Abiding by the Laws), *Zhongguo huanjing bao* (*China Environment News*), August 4, 2017. www.cenews.com.cn/syyw/201708/t20170804_845266.html.
127. "Jingjinji da ducha: qinchai henmang, jiceng jinzhang" ("Supervision in Beijing-Tianjin-Hebei: Imperial Envoy is Very Busy, and the Grassroots Cadres are Nervous"), *Caixin zhoukan* (*Caixin Weekly*), September 1, 2017. http://china.caixin.com/2017-09-01/101138863.html.
128. The Ministry of Environmental Protection, "Zhengzhi sanluanwu liuwan jia" (Renovating 60,000 "'Scattered, Chaotic and Polluted' enterprises"), *Beijing wanbao* (*Beijing Evening News*), January 18, 2018. http://mp.163.com/v2/article/detail/D8EB37SM0514C7JA.html.
129. Wu Nan, "Farewell to 'Apec Blue' as Smog Returns to Beijing Following Summit," *South China Morning Post*, November 19, 2014. www.scmp.com/news/china/article/1643740/farewell-apec-blue-smog-returns-beijing-following-summit.
130. Haoran Liu, Cheng Liu, Zhouging Xie, et al., "A Paradox for Air Pollution Controlling in China Revealed by 'APEC Blue' and 'Parade Blue'," *Scientific Reports* 6, no. 34408 (September, 2016), 1–13.
131. Peking University Statistical Science Center and Peking University Guanghua School of Management, *Kongqi zhiliang pinggu baogao (Air Quality Assessment Report)*, March 2015.
132. Anna Fifield, "Beijing's Smog Could Be Party Crasher for Giant 70-year Celebrations," *The Washington Post*, September 27, 2019.
133. Shi Qingling, Guo Feng, and Chen Shiyi, "Wumai zhili zhong de zhengzhi xing lantian" ("Political Blue Sky in Smog Control"), *Zhongguo gongye jingji* (*China Industrial Economics*), no. 5, 2016.

134. Serenitie Wang and Katie Hunt, "China: 'Political' Blue Sky Comes at a Price," *CNN*, December 15, 2016. www.cnn.com/2016/12/15/asia/china-air-pollution-study/index.html.
135. Chang Jiwen, "Ruhe pojie huanjing baohu de xingshizhuyi" ("How to Address Formalism in Environmental Protection"), Research Center of the State Council, January 5, 2018. www.chinareform.net/index.php?m=content&c=index&a=show&catid=319&id=23776
136. Chinese scholar, private communication with author, August 14, 2018.
137. Zhang Ke, "Difang jinji tingchan tingye yingdui ducha" ("Local Production is Suspended to Cope with Supervision"), *Diyi caijing (yicai)*, December 28, 2017. www.yicai.com/news/5387338.html.
138. "17.6 wan jia zhizao qiye tinggong" ("176,000 Factories Suspended Operation"), December 16, 2017. http://wemedia.ifeng.com/41223113/wemedia.shtml.
139. Cao Hongyan, "Huanbaobu: jingjinji quye 17.6 wan jia sanluanwu qiye jiu yue di qian xu dabiao paifang" ("176,000 Scattered, Chaotic and Polluted Enterprises in Beijing-Tianjin-Hebei Area are to Meet Standards before Discharging Pollutants"), *zhongguo jingji wang*, July 14, 2017. www.ce.cn/xwzx/gnsz/gdxw/201707/14/t20170714_24228961.shtml.
140. David Stanway, "China Environment Facing Pressures from Economic Slowdown – Minister," *Reuters*, October 29, 2018.
141. Zhang Ke, "Zhongyang jingji gongzuo huiyi: dahao wuran fangzhi gongjianzhan bimian chuzhi cuoshi jiandan cubao" ("Central Economic Work Conference: Do a Good Job in Pollution Prevention and Control, and Avoid Oversimplified and Crude Measures"), *Diyi caijing (yicai)*, December 21, 2018. www.yicai.com/news/100084815.html.
142. Ministry of Ecology and Environment, "Bei huanbaobu tongbao gao huanbao yidaoqie Shandong linyi lianye kaihui" ("Linyi of Shandong Held an Overnight Meeting after Being Publicly Names for its Cookie-cutter Approach to Environmental Protection"), *Xin jing bao* (*Beijing News*), September 5, 2019. https://news.sina.com.cn/c/2019-09-05/doc-iicezueu3705136.shtml.
143. "Qi changwei qunian 560 ci pishi huanbao" ("Seven Politiburo Standing Committee Members 560 Times Provided Written Comments on Environmental Protection"), *Beijing qingnian bao* (*Beijing Youth Daily*), January 16, 2015. http://news.sohu.com/20150116/n407832428.shtml.
144. Liu Shizhen, "Huanbaobu zoufang wubai yu wan hu" ("MEP Interviewed More than Five Million Households"), *Zhongqing zaixian*, December 24, 2017. www.chinanews.com/gn/2017/12-24/8408014.shtml.
145. "Jingjinji 28 cheng wancheng meigai jin 400 wan hu" ("28 Cities in Beijing-Tianjin-Hebei Region Completed the Conversion of Four Million Households"), *caixin wang*, December 26, 2017. www.chinaiol.com/heating/r/1226/12190731.html.
146. "Jingjinji 28 cheng wancheng meigai jin 400 wan hu" ("28 Cities in Beijing-Tianjin-Hebei Region Completed the Conversion of Four Million Households").
147. Joe McDonald, "China Suffers Natural Gas Shortage as Coal Ban Backfires," *Associated Press*, December 14, 2017. www.apnews.com/9c1d7414c446464c8c3aaaaa183abf27

148. Xiao Yan et al., "Emission Characteristics of Gas-Fired Boilers in Beijing City, China: Category-Specific Emission Factor, Emission Inventory, and Spatial Characteristics," *Aerosol and Air Quality Research*, 17 (2017): 1825–183.
149. Wang Fan and Li Junfeng, "Zhimai buneng paopian" ("Smog Control Should Not Be Off Track"), *Nengyuan pinglun* (*The Energy Review*), no. 4, 2014.
150. Keith Crane and Zhimin Mao, *Costs of Selected Policies to Address Air Pollution in China* (Santa Monica, CA: RAND Corporation, 2015), 3. www.rand.org/pubs/research_reports/RR861.html.
151. McDonald, "China Suffers Natural Gas Shortage as Coal Ban Backfires."
152. Zhuang Jian, "Fupan qihuang Yichang yibei yujian de gongxu shiheng" ("A Second Count of Gas Shortage: An Anticipated Imbalance between Supply and Demand), *yuandongli*, December 20, 2017. https://m.jiemian.com/article/1830430.html
153. "Mei gai qi bei huanbaobu jinji jiaoting" ("Conversion from Coal to Gas Was Abruptly Stopped by MEP"), December 3, 2018. www.sohu.com/a/279408055_617145
154. "Ganga le: meigaiqi wancheng, kaishi gongnuan le, qi que mei le" ("An Awkward Situation: The Conversion from Coal to Gas Heating is Complete and the Heating is On, But the Gas is Not Available"), November 22, 2017. www.china5e.com/news/news-1011010-1.html.
155. McDonald, "China Suffers Natural Gas Shortage as Coal Ban Backfires."
156. Liu Shizhen, "Huanbaobu zoufang wubai yu wan hu" ("MEP Interviewed More Than Five Million Households"), *Zhongqing zaixian*, December 24, 2017. www.chinanews.com/gn/2017/12-24/8408014.shtml.
157. McDonald, "China Suffers Natural Gas Shortage as Coal Ban Backfires."
158. Alexander Mercouris, "China Finally Switching to Russian Gas," *The Duran*, February 11, 2018. http://theduran.com/china-disillusioned-central-asian-gas-switches-russian-gas/ ; Jin Canrong, Remarks at Expert Salon Meeting Summary on U.S.-China Trade War, April 2, 2018.
159. Viola Zhou, "China U-turns on Coal Ban Amid Growing Outcry over Numbers Left Freezing in Winter Cold," *South China Morning Post*, December 7, 2017. www.scmp.com/news/china/policies-politics/article/2123270/china-u-turns-coal-ban-amid-growing-outcry-over-numbers.
160. "Guojia fazhan gaigewei bangongtin guojia nengyuanju zonghesi guanyu jiada qingjiemei gongying quebao qunzhong wennuan guodong de tongzhi" ("Circular of NDRC Office and NEA Comprehensive Division on Scaling Up Clean Coal Supply to Keep People Warn in Winter"), National Development and Reform Commission Office, December 25, 2017. www.ndrc.gov.cn/zcfb/zcfbtz/201712/t20171226_871677.html.
161. "Shanxi Taiyuan yidaoqie jin yong mei" (Taiyuan of Shanxi Adopts a Cookie-cutter Approach, Prohibiting Coal Use"), *Renmin ribao (People's Daily)*, November 17, 2018. https://news.163.com/18/1117/17/E0R4IESU0001875P.html.
162. Phoebe Zhang, "Chinese City to Investigate after Six People Die from Carbon Monoxide Poisoning after Burning 'Clean Coal'," *South China Morning Post*, December 5, 2019. www.scmp.com/news/china/society/article/3040755/chinese-city-investigate-after-six-people-die-carbon-monoxide.

163. Wang Wenting, "Lingyu faxue shiye xia pojie daqi wuran weiji de caishuifa zhi dao" ("Finance and Tax Law Solution for Air Pollution Crisis from a Realm Law Perspective"), *Caishuifa luncong* (*Finance and Tax Law Review*), no. 17, 2018.
164. Liu Shuyan, *Dangdai zhongguo zhengfu yu zhengzhi* (*Contemporary Chinese Government and Politics*) (Beijing: Zhongguo shehui kexue chubanshe, 2016).
165. Lan Xue and Jing Zhao, "Truncated Decision Making and Deliberate Implementation: A Time-Based Policy Process Model for Transitional China," *Policy Studies Journal* 48, no. 2 (May, 2020), 298–326.

CONCLUSION

1. Claire Che, "Beijing's Air Quality is Worse Than Smoke-Filled California Cities," *Bloomberg News*, November 14, 2018. www.bloomberg.com/news/articles/2018–11-14/beijing-s-worst-air-in-more-than-year-engulfs-buildings-in-smog.
2. "Beijing qian shiyue PM2.5 nongdu chuang lishi zuidi" (Beijing's Average PM2.5 Concentration over the Pasts 10 Months Dropped to a Historical Low), *Beijing qingnian bao (Beijing Youth Daily)*, November 16, 2018, A4.
3. James Fallows, "2 Charts That Put the Chinese Pollution Crisis in Perspective," *The Atlantic*, April 18, 2014. www.theatlantic.com/international/archive/2014/04/2-charts-that-put-the-chinese-pollution-problem-in-perspective/360868/.
4. "Xinlang weibo shang de yan xuetong" ("Yan Xuetong" on Sino weibo), https://free weibo.com/weibo/%E9%98%8E%E5%AD%A6%E9%80%9A?latest.
5. John Mearsheimer's interview with *Nanfang dushi bao* (*Southern Metropolis Daily*), June 8, 2012; Richard Haas, Remarks at CFR Luncheon at ISA: A World in Disarray, San Francisco, April 6, 2018.
6. Jessica Chen Weiss, "A World Safe for Autocracy? China's Rise and the Future of Global Politics," *Foreign Affairs*, (July/August 2019).
7. Yan Xuetong, *Leadership and the Rise of Great Powers* (Princeton, NJ: Princeton University Press, 2019).
8. Maria Abi-Habib and Hari Kumar, "India Finally Has Plan to Fight Air Pollution. Environmentalists Are Wary." *New York Times*, January 11, 2019. www.nytimes.com/2019/01/11/world/asia/india-air-pollution.html.
9. Ministry of Environmental Protection, "Daqi wuran fangzhi meiti jianmianhui" ("Press Conference on Air Pollution Control"), January 7, 2017. http://m.news.cctv.com/2017/01/07/ARTIVrid1dE4q8k2nOQsrRYX170107.shtml.
10. Steven Ratner, "Is China's version of capitalism winning?" *New York Times*, March 27, 2018.
11. See Joseph Fewsmith and Andrew J. Nathan, "Authoritarian Resilience Revisited: Joseph Fewsmith with Response from Andrew J. Nathan," *Journal of Contemporary China* 28, no. 116 (2019): 167–179.
12. Tyrene White, "Postrevolutionary Mobilization in China: The One-Child Policy Reconsidered," *World Politics* 43 (October, 1990): 53–76. See also Zheng Wang, "Patriotic Education Campaign in China," *International Studies Quarterly* 52, no. 4 (December, 2008): 783–806.

13. Genia Kostka and Chunman Zhang, "Tightening the Grip: Environmental Governance under Xi Jinping," *Environmental Politics* 27, no. 5 (2018): 778.
14. George Rosen, "Political Order and Human Health in Jeffersonian Thought," *Bulletin of the History of Medicine* 26 (1952): 32–44.
15. Thomas Bollyky et al., "The Relationships between Democratic Experience, Adult Health, and Cause-specific Mortality in 170 Countries between 1980 and 2016: An Observational Analysis," *Lancet* 393, no. 10181 (April, 2019): 1628–1640.
16. On the need of such transition, see Ran Ran, *Zhongguo defang huanjing zhengzhi: zhengce yhu zhixing zhijian de juli (China's Local Environmental Politics: The Distance between Policy and Enforcement)* (Beijing: Zhongyang bianyi chubanshe, 2015).
17. Michel Foucault, "Governmentality," (trans.) Rosi Braidotti and revised by Colin Gordon, in Graham Burchell, Colin Gordon and Peter Miller (eds.) *The Foucault Effect: Studies in Governmentality* (Chicago, IL: University of Chicago Press, 1991), 87–104.
18. Carl Minzner, *End of An Era: How China's Authoritarian Revival is Undermining Its Rise* (New York: Oxford University Press, 2018).
19. Benjamin L. Liebman and Curtis J. Milhaupt, *Regulating the Visible Hand? The Institutional Implications of Chinese State Capitalism* (New York: Oxford University Press, 2015), xxv.
20. Stein Ringen, *The Perfect Dictatorship: China in the 21st Century* (Hong Kong: HKU Press, 2016), p. ix.
21. In a major speech he delivered in October 2019, Xi reiterated that rule of law, orderly succession of state leadership, and broad public participation in political life are crucial criteria in judging whether a country's political system is democratic or effective. See Xi Jinping, "Jianchi he wanshan zhongguo tese shehui zhuyi zhidu tuijing guojia zhili tixi he zhili nengli xiandaihua" ("Upholding and Improving the System of Socialism with Chinese Characteristics and Promoting the Modernization of State Governance System and Governance Capacity"), *Qiushi* (*Seek Truth*), January 2, 2020.
22. Ian Johnson, "How the Communist Party Guided China to Success," *New York Times*, February 22, 2017. www.nytimes.com/2017/02/22/world/asia/china-politics-xi-jinping.html; Minxin Pei, "China's Coming Upheaval: Competition, the Coronavirus, and the Weakness of Xi Jinping," *Foreign Affairs*, May/June 2020. https://www.foreignaffairs.com/articles/united-states/2020-04-03/chinas-coming-upheaval.
23. Michael Beckley, "The United States Should Fear a Faltering China," *Foreign Affairs* website, October 28, 2019. www.foreignaffairs.com/articles/china/2019–10-28/united-states-should-fear-faltering-china.

Index

CPSIA information can be obtained
at www.ICGtesting.com
Printed in the USA
LVHW092014111220
673944LV00006B/896